London Mathematical Society Lecture Note Series 27

Skew Field Constructions

P. M. COHN

Bedford College
University of London

CAMBRIDGE UNIVERSITY PRESS
CAMBRIDGE
LONDON NEW YORK MELBOURNE

Published by the Syndics of the Cambridge University Press
The Pitt Building, Trumpington Street, Cambridge CB2 1RP
Bentley House, 200 Euston Road, London NW1 2DB
32 East 57th Street, New York, NY 10022, USA
296 Beaconsfield Parade, Middle Park, Melbourne 3206, Austral.

First published 1977

Printed in Great Britain
at the University Press, Cambridge

Library of Congress Cataloguing in Publication Data

Cohn, Paul Moritz.

Skew field constructions.

(London Mathematical Society lecture note series; 27)
Bibliography
Includes index.
1. Division rings. 2. Fields, Algebraic. I. Title.
II. Series: London Mathematical Society. Lecture note series; 27.
QA251.5.C63 1977 512'.4 76-46854

ISBN 0 521 21497 1

Contents

Preface

The history of skew fields begins with quaternions, whose discovery (in 1843) W.R. Hamilton regarded as the climax of his far from ordinary career. But for a coherent theory one has to wait for the development of linear associative algebras; in fact it was not until the 1930's that a really comprehensive treatment of skew fields (by Hasse, Brauer, Noether and Albert) appeared. It is an essential limitation of this theory that only skew fields finite-dimensional over their centres are considered.

Although general skew fields have made an occasional appearance in the literature, especially in connexion with the foundations of geometry, very little of their properties was known until recently, and even particular examples were not easy to come by. The first well known case is the field of skew power series used by Hilbert in 1898 to illustrate the fact that a non-archimedean ordered field need not be commutative. There are isolated papers in the 1930's, 1940's and 1960's (Moufang, Malcev, B.H. Neumann, Amitsur and the author) showing that the free algebra can be embedded in a skew field, but the development of the subject is hampered by the fact that one has no operation that can be performed on skew fields (over a given ground field) and again produces a skew field. In the commutative case one has the tensor product, which leads to a ring, from which fields can then be obtained as homomorphic images. The corresponding object in the general case is the free product and in the late 50's the author tried to prove that this could be embedded in a skew field. This led to the development of firs (= free ideal rings); it could be shown (1963) that any free product of skew fields is a fir, but it was not until 1971 that the original aim was achieved, by proving that every fir is embeddable in a skew field, and in fact has a universal field of fractions. Combining these results, one finds that any free product or 'coproduct' of fields has a universal field of fractions, or a field coproduct, as we shall call it. It is this result which forms the starting point for these lecture notes.

As the above description shows, the proof of the existence of field coproducts falls into two parts, showing (a) that the coproduct of skew fields is a fir, and (b) that every fir has a universal field of fractions. Of these, (b) was proved and discussed very fully in the author's 'Free Rings'; for this reason the account given below (in Ch.4) leaves out some of the longer proofs. Thus the chapter can serve as a reminder of, replacement for or introduction to 'Free Rings', as the case may be. The result (a) was first proved by the author in 1963, but in 1974 Bergman gave a very far-reaching generalization, and it is Bergman's results which we present here, in Ch.5, in a simplified form, for which I am indebted to W. Dicks. Ch.5 also contains some applications, including a simple example of a skew field extension with different left and right degrees (Artin's problem). The construction of Ch.4 allows us to give a general discussion of skew field extensions in terms of presentations (Ch.6), including the notion of algebraic closure and the word problem: we shall see that the word problem for free fields can be solved (at least in a relative sense) and also meet a simple example, due to Macintyre, of a skew field with unsolvable word problem.

One of the central problems of skew field theory is the solving of equations, and this makes it important to have a good notion of specialization. Unfortunately specialization in skew fields lacks many of the good features we are accustomed to, and very little is known about it so far. However a recent study by Bergman (in which he uses his work with Small on polynomial identities) has led to some remarkable and surprisingly detailed results on the structure of specializations, and these form the content of Ch.7; I am grateful to G.M. Bergman for allowing me to include this material before publication.

In Ch.8 we come to the actual solving of equations; the results so far are quite meagre, but it seemed worthwhile at least to set out the problems in what may be a more accessible form. The main results include the similarity reduction of a matrix over a skew field, and the reduction of the main problem (solving equations) to an eigenvalue problem. Some of these results have not appeared in print before, but even for those that have, it was usually possible to give a simpler presentation here, in the framework of a general development of the subject.

The first three chapters deal with matters which are better known (though not all have found their way into text-books yet), partly to show their relation to the rest of the notes but also to make the notes more widely accessible. Apart from Ore's method of skew polynomials (Ch.1) and skew power series (Ch.2) it includes a discussion of extensions of finite degree (Ch.3), in particular Amitsur's theory of cyclic extensions, and an account of Galois theory. No special prerequisites are needed, beyond a standard algebra course.

The bibliography includes (besides the works referred to in the text) a number of papers on skew fields infinite-dimensional over their centres, but does not lay claim to completeness.

As the name indicates, these really are lecture notes, though not for a single set of lectures. For this reason they may lack the polish of a book, but it is hoped that they have not entirely lost the directness of a lecture. The material comes from courses I have given in Manchester and London; some parts follow rather closely lectures given at Tulane Universit (1971), the University of Alberta (1972), Carleton University (1973), Tübingen (1974), Mons (1974), Haifa Technion (1975), Utrecht (1975) and Ghent (1976). It is a pleasure to acknowledge the hospitality of these institutions, and the stimulating effect of such critical audiences.

In preparing these notes I have greatly benefitted from the help of Dr. Warren Dicks: his careful reading of the manuscript led to the removal of many obscurities and errors; he has also read the proofs and I am most grateful for all his help. I should also like to thank both the staff of the Cambridge University Press and Margaret Harrison for the enthusiastic and efficient way in which they have coped with a difficult manuscript.

P.M. Cohn

Bedford College
London
December 1976

Prologue

O glücklich, wer noch hoffen kann,
Aus diesem Meer des Irrtums aufzutauchen!
Was man nicht weiss, dass eben brauchte man,
Und was man weiss, kann man nicht brauchen.

Goethe, Faust I

The object of these notes is to describe some methods of constructing skew fields. The case most studied so far is that of skew fields finite-dimensional over their centres. But a finite-dimensional k-algebra (where k is a commutative field) is a field whenever it has no zero-divisors. On the one hand this enormously simplifies the study, while on the other hand it puts many constructions out of bounds (because they produce infinite-dimensional algebras). The study of fields that are not necessarily finite-dimensional over their centres is still in its early stages, and the methods needed here are not very closely related to those used on finite-dimensional algebras, - the relation between these subjects is rather like the relation between finite and infinite groups.

There are some ways of obtaining a field directly, e.g. Schur's lemma tells us that the centralizer of a simple module is a field, and the coordinatization theorem shows that when we coordinatize a desarguesian plane, the coordinates lie in a field. But these methods are not very explicit, and we shall have no more to say about them. For us the usual way to construct a field is to take a suitable ring and embed it in a field. What is to be understood by 'suitable' will transpire later.

There are four methods of interest to us; they are

(1) Ore's method (Ch.1),

(2) The method of power series (Ch.2),

(3) Inverse limits of Ore rings (Ch.2),

(4) A general criterion (Ch.4).

As a test ring we shall use the free algebra on a set X over a commutative field k, written k<X>. All four methods can be used on k<X>, and each has its pros and cons. (1) is particularly simple, (2) and (3) provide a convenient normal form, while (4) gives (at least in principle) a complete survey over all possible embeddings, indeed over all homomorphisms of k<X> into fields.

As the reader will have noticed, we use the term 'skew field' to mean an associative but not necessarily commutative division ring. Frequently we omit the 'skew', which means that we must qualify the commutative fields. We shall also assume that every ring has a unit-element, denoted by 1, which is inherited by subrings, preserved by homomorphisms and acts unitally on all modules. We note that in a ring $1 = 0$ is possible; it happens precisely when the ring is *trivial* i.e. it reduces to 0. A ring is said to be *entire* or an *integral domain* if $1 \neq 0$ and the product of non-zero elements is non-zero. In any ring R, the set of non-zero elements is denoted by R*.

As usual, **N**, **Z**, **Q**, **R**, **C** stand for the natural numbers, integers, rational numbers, real numbers and complex numbers respectively. If K is a ring, then by a K-*algebra* we understand a ring R with a given homomorphism of K into the centre of R, and a K-*ring* R is a ring R with a homomorphism of K into R; this agrees with the usual definition, because R has a 1.

1·Skew polynomial rings and their fields of fractions

1.1 The general embedding problem

Let R be a ring; by a *field of fractions* of R we understand a field K together with an embedding $R \longrightarrow K$ such that K is the field generated by the image of R. Our task then is to find when a ring has a field of fractions. For commutative rings the answer is easy (and well known). It falls into three parts:

(i) Existence. *A field of fractions exists for a ring* R *if and only if* R *is an integral domain.*

(ii) Uniqueness. *When the field of fractions exists, it is unique up to isomorphism, i.e. given any two fields of fractions of* R, $\lambda_i : R \longrightarrow K_i$ $(i = 1,2)$, *there exists an isomorphism* $\phi : K_1 \longrightarrow K_2$ *such that the diagram shown commutes.*

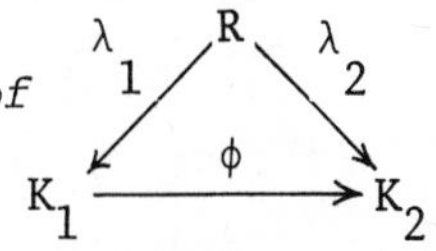

$$\begin{array}{ccc} & R & \\ {}^{\lambda_1}\swarrow & & \searrow{}^{\lambda_2} \\ K_1 & \xrightarrow{\ \phi\ } & K_2 \end{array}$$

(iii) Normal form. *Each element of the field of fractions can be written in the form* a/b, *where* $a,b \in R$ $(b \neq 0)$ *and* $a/b = a'/b'$ *if and only if* $ab' = ba'$.

Of course this is not really a 'normal form'; only in certain rings such as **Z** and $k[x]$ is there a canonical representative for each fraction.

Let us now pass to the non-commutative case. The absence of zero-divisors is still necessary for a field of fractions to exist, but not sufficient. The first example is due to Mal'cev [37] who writes down a semigroup whose semigroup ring over **Z** is entire but is not embeddable in a field. To describe his example we take 8 generators a_{ij}, b_{ij} $(i,j = 1,2)$ written in matrix form as $A = (a_{ij})$, $B = (b_{ij})$, with defining relations (again in matrix form)

$$AB = \begin{pmatrix} 0 & c \\ 0 & 0 \end{pmatrix},$$

where c is arbitrary. In other words, we have the three relations expressing that the (1,1), (2,1) and (2,2)-entries of AB vanish. Then c, the (1,2)-entry of AB, as well as the entries of A, B are different from 0. Now if R were embeddable in a field, we could reduce the (1,1)-entry of A to 0 by subtracting a suitable multiple of row_2 from row_1; this leaves the product AB unchanged. Similarly we can reduce the (2,2)-entry of B to 0 by subtracting a multiple of col_1 from col_2; again this has no effect on AB. We now have

$$(1) \qquad \begin{pmatrix} 0 & * \\ * & * \end{pmatrix} \begin{pmatrix} * & * \\ * & 0 \end{pmatrix} = \begin{pmatrix} 0 & c \\ 0 & 0 \end{pmatrix},$$

which is a contradiction, because $c \neq 0$.

Thus R has no field of fractions. To prove that R is entire Mal'cev uses a normal form argument, but there is a more illuminating way of seeing this, which will help to explain some of the conditions found in Ch.4. We shall need to anticipate some of the definitions given there.

A relation

$$(2) \qquad x_1y_1 + \ldots + x_ny_n = 0$$

in any ring R is called *trivial* if for each $i = 1,\ldots,n$, either $x_i = 0$ or $y_i = 0$. Every ring $\neq \{0\}$ has non-trivial relations, e.g. $(1,1)\begin{pmatrix} 1 \\ -1 \end{pmatrix} = 0$, but this relation still has an air of triviality about it. To make this precise, let us call a relation $x.y = 0$ between a row x and a column

y *trivializable* if there exists an invertible matrix P over R such that $xP.P^{-1}y = 0$ is a trivial relation. There are rings in which every relation is trivializable, e.g. $\mathbf{Z}$ or $k[x]$; on the other hand, in $\mathbf{Z}[x]$ the relation $2.x - x.2 = 0$ is not trivializable, neither is $x.y - y.x = 0$ in $k[x,y]$. So the following definition makes sense.

A non-trivial ring in which every relation of at most n terms (as in (2)) is trivializable is called an n-*fir* (fir = free ideal ring, cf. 4.3). An n-fir can also be characterized by the property that every n-generator right ideal is free (as module over the ring) of unique rank. A 1-fir is just an entire ring, as is easily checked. It is clear that every n-fir is an (n-1)-fir, so for increasing n, the class of n-firs becomes smaller (in fact the inclusions are strict at each stage, cf.Cohn [69]). A ring which is an n-fir for each n is called a *semifir*; this then means that each finitely generated right ideal is free, of unique rank, and by symmetry the same holds for left ideals. In particular, a commutative semifir is just a Bezout domain (all finitely generated ideals are principal), so in the commutative case a 2-fir is already a semifir.

Now there is a result which states that if a ring R has a presentation whose relations each have more than n terms (for some fixed n) of sufficiently nice form, to prevent interaction (of course this must be made precise), then R is an n-fir (cf.Cohn [69]). In particular, Mal'cev's ring is defined by 2-term relations $a_{11}b_{11} + a_{12}b_{21} = 0$, etc. and these satisfy the conditions, so it is a 1-fir, i.e. entire.

Mal'cev expressed his example as a cancellation semigroup not embeddable in a group, and it prompted him to ask for a ring R whose set R* of non-zero elements can be embedded in a group but which cannot itself be embedded in a field. This question was answered affirmatively 30 years later, in 1966, by three people simultaneously and independently: Bokut' [69],

Bowtell [67] and Klein [67]. We shall return to this problem in 4.3.

Mal'cev, after giving his example, went on in a remarkable pair of papers (Mal'cev [39]) to provide a set of necessary and sufficient conditions for a semigroup to be embeddable in a group. This is an infinite set of conditions, and Mal'cev showed that no finite subset could be sufficient. The first conditions expressed cancellation, then came the condition already encountered (expressing that c in (1) must be 0), and the other conditions were similar but more complicated (cf. Mal'cev [39], Cohn [65]). All were of the form

$$A_1 \wedge A_2 \wedge \ldots \wedge A_n \Rightarrow B, \tag{3}$$

where $A_1,\ldots,A_n,B$ are certain equations, with universal quantifiers for all the variables prefixed. Such a condition (3) is called a *quasi-identity* (for an identity the A's are missing). As a matter of fact it follows from general principles of universal algebra that the class of semigroups embeddable in grou is a *quasi-variety*, i.e. definable by quasi-identities. For i can be shown to be a universal class (i.e. definable by sentences with universal quantifiers only) and one has the following theorem (Cohn [65], p.235):

A universal class of algebras is a quasi-variety if and onl if it admits direct products and contains the 1-*element algebr*

With this result it is not hard to check that the class of semigroups embeddable in groups forms a quasi-variety. At the same time we see that entire rings do not, since they do not admit direct products, and neither do rings embeddable in fields. Nevertheless they come very close to being a quasi-variety; to be precise, if $\mathcal{D}$ is the class of entire rings, the there is a quasi-variety $\mathcal{R}$ of rings such that $\mathcal{D} \cap \mathcal{R}$ is the exact class of rings embeddable in fields. To obtain $\mathcal{R}$, we recall a definition: A ring R is said to be *strongly regular* if for each $x \in R$ there exists $y \in R$ such that $x^2y = x$ (one can show that this is stronger than v.Neumann-regularity; a ring is strongly regular if and only if it is regular and has no non-zero nilpotent elements). Now one has the following facts, which enable us to identify $\mathcal{R}$ as the class of rings embeddable in strongly regular rings:

(a) *The class of subrings of strongly regular rings is a quasi-variety,*

(b) *an entire ring is embeddable in a strongly regular ring if and only if it is embeddable in a field.*

Of these (a) is easily checked by the criterion given earlier; (b) is proved by showing that every strongly regular ring is a subdirect product of fields (cf. e.g. Cohn [71']). It follows that

(c) *The class of entire rings embeddable in fields can be defined by quasi-identities.*

Mal'cev's conditions for embedding semigroups in groups have been used in a most ingenious way by A. Klein [69] to obtain the following result:

Let R be an entire ring satisfying the following condition for all n:

(4) *If* C *is a nilpotent* n x n *matrix over* R, *then* $C^n = 0$,

then R* can be embedded in a group.

Klein also showed that his condition (4) is not necessary for R* to be embeddable in a group; but it is clearly necessary for R to be embeddable in a field. Condition (4) for entire rings implies most known conditions for a field of fractions to exist such as invariant basis number (Klein [70,72]), but it is not sufficient, as examples by Bergman [74] show.

1.2 Ore's method

We shall now treat a special case where the field of fractions always exists and is particularly simple to construct. But first let us look at the problem of constructing inverses quite generally.

Let R be a ring and S a subset of R. A homomorphism $f:R \longrightarrow R'$ to another ring R' is called S-*inverting* if for each $s \in S$, sf is an invertible element of R' (i.e. an element with a two-sided inverse, also called a *unit*). The following result, although trivial to prove, is useful in considering S-inverting maps.

Proposition 1.2.1. *Given a ring* R *and a subset* S *of* R, *there exists a ring* R_S *and an* S-*inverting homomorphism* $\lambda:R \longrightarrow R_S$ *which is universal* S-*inverting, in the sense that for each* S-*inverting homomorphism* $f:R \longrightarrow R'$ *there is a unique homomorphism* $f':R_S \longrightarrow R'$ *such that* $f=\lambda f'$.

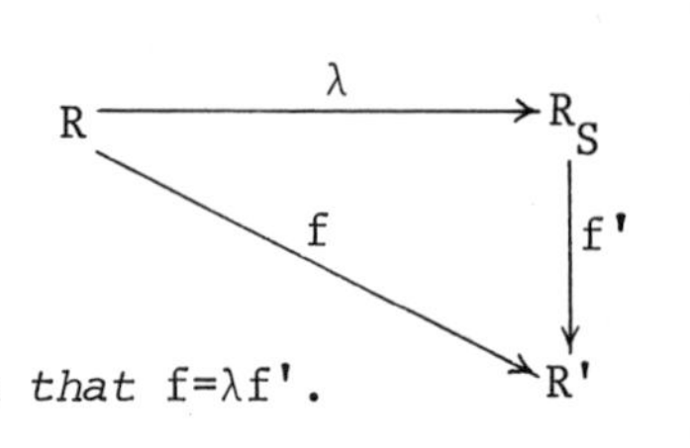

As with all universal constructions, the universal property determines R_S up to isomorphism.

To prove the existence of R_S we take a presentation of R and for each $s \in S$ adjoin an element s' with defining relations $ss' = s's = 1$. The map λ is defined by assigning to each element in R the corresponding element in the given presentation. Then sλ is invertible, the inverse being s'. Thus we have a ring R_S and an S-inverting map $\lambda:R \longrightarrow R_S$. Given any S-inverting homomorphism $f:R \longrightarrow R'$, we define $f':R_S \longrightarrow R'$ by mapping aλ to af ($a \in R$) and s' to $(sf)^{-1}$, which exists in R', by hypothesis. Any relation in R_S must be a consequence of relations in R and relations expressing that s' is the inverse of sλ. All these relations still hold in R', so f' is well-defined and it is clearly a homomorphism. It is unique because its values on Rλ are prescribed, as well as on $(S\lambda)^{-1}$, by the uniqueness of inverses. ■

The ring R_S constructed here is called the *universal S-inverting ring* for the pair R,S. We have in fact a functor from pairs (R,S) consisting of a ring R and a subset S of R (with morphisms $f:(R,S) \longrightarrow (R',S')$ homomorphisms from R to R' which map S into S') to the category of rings and homomorphisms. All this is easily checked, but it provides no information about the structure of R_S. In particular we shall be interested in a normal form for the elements of R_S and an indication of the size of the kernel of λ, and here we shall need to make some simplifying assumptions.

Let us look at the commutative case first. To get a convenient expression for the elements of R_S we shall take S to be *multiplicative*, i.e. $1 \varepsilon S$ and $a,b \varepsilon S \Rightarrow ab \varepsilon S$. Then every element of R_S can be written as a quotient a/s, where $a \varepsilon R$, $s \varepsilon S$, and $a/s = a'/s'$ if and only if $as't = a'st$ for some $t \varepsilon S$. This is not exactly what one understands by a normal form, but it is sufficiently explicit to allow us to determine the kernel of λ, viz.

(1) $\ker \lambda = \{a \varepsilon R \mid at = 0 \text{ for some } t \varepsilon S\}$.

Ore's idea consists in asking under what circumstances the elements of R_S have this form, when commutativity is not assumed. We must be able to express $s^{-1}a$ (for $a \varepsilon R$, $s \varepsilon S$) as a_1/s_1, where $a_1 \varepsilon R$, $s_1 \varepsilon S$, and multiplying up, we find $as_1 = sa_1$. More precisely, we have $as_1/1 = sa_1/1$, whence $as_1t = sa_1t$ for some $t \varepsilon S$. This is the well known *Ore condition* and it leads to the following result:

Theorem 1.2.2. *Let* R *be a ring and* S *a subset such that*

D.1 S *is multiplicative,*

D.2 *For any* $a \varepsilon R$, $s \varepsilon S$, $sR \cap aS \neq \emptyset$,

D.3 *For any* $a \in R$, $s \in S$, $sa = 0 \Rightarrow at = 0$ *for some* $t \in S$. *Then the universal* S-*inverting ring* R_S *may be constructed as follows: On* $R \times S$ *define the relation*

(2) $(a,s) \sim (a',s')$ *whenever* $au = a'u'$, $su = s'u' \in S$ *for some* $u,u' \in R$.

This is an equivalence on $R \times S$ *and the quotient* $R \times S/\sim$ *is* R_S. *In particular, the elements of* R_S *may be written as fractions* $a/s = as^{-1}$ *and* $\ker \lambda$ *is given by* (1).

The proof is a lengthy but straightforward verification, which may be left to the reader. It can be simplified a little by observing that the assertion may be treated as a result on semigroups; once the 'universal S-inverting semigroup' R_S has been constructed by the method of this theorem, it is easy to extend the ring structure to R_S. A subset S of a ring R satisfying D.1-3 is called a *right denominator set*.

When R is commutative, D.2-3 are automatic and may be omitted. If R is entire, D.3 may be omitted and if moreover, $S = R^*$, then D.2 reads $aR \cap bR \neq 0$ for $a,b \neq 0$. This was the case actually treated by Ore [31] and R is then called a *right Ore domain*. Apparently these results were found independently by E. Noether, but not published. There have been many papers dealing with generalizations, e.g. Asano [49]; for a survey see Cohn [71'].

It is important to observe that the field of fractions of a right Ore domain is essentially unique. Let us first note that the construction is functorial. Thus, given a map between pairs $f:(R,S) \longrightarrow (R',S')$, i.e. a homomorphism $f:R \longrightarrow R'$ such that $Sf \subseteq S'$, then we have the diagram shown, and by universality of R_S there is a unique map $f_1:R_S \longrightarrow R'_{S'}$, such that the resulting square commutes. In particular, if f is an isomorphism, so is f_1.

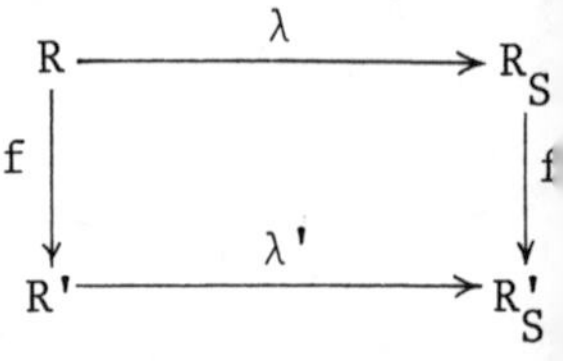

$$\begin{array}{ccc} R & \xrightarrow{\lambda} & R_S \\ {\scriptstyle f}\downarrow & & \downarrow{\scriptstyle f_1} \\ R' & \xrightarrow{\lambda'} & R'_{S'} \end{array}$$

So far R,R' have been quite general; suppose now that R is a right Ore domain and K is any field of fractions of R, thus we have an embedding $f:R \longrightarrow K$. If $S = R^*$, we have a homomorphism $f_1:R_S \longrightarrow K$, which we claim is injective. For if $as^{-1} \varepsilon \ker f_1$, then $0 = (as^{-1})f_1 = (af)(sf)^{-1}$, hence $af = 0$, and so $a = 0$, because f is injective. It follows that f_1 is an embedding; the image is a field containing R and hence equal to K, because K was a field of fractions. Thus f_1 is an isomorphism and we have proved

Proposition 1.2.3. *The field of fractions of a right Ore domain is unique up to isomorphism.* ■

The result is of particular interest because it ceases to hold for more general rings; we shall soon meet rings which have several non-isomorphic fields of fractions.

1.3 Skew polynomial rings

Given a commutative field k, there are four important constructions involving an indeterminate that we can perform. We can form polynomials, rational functions, formal power series and formal Laurent series. These constructions are all well known in the commutative case, and the relations between the four rings so obtained may be summed up in the following commutative diagram

Each can be generalized by taking the field k to be skew, and taking the indeterminate x to be central, but that is not the most general (nor the most useful) choice.

Starting from an entire ring A, let us ask for a ring R whose elements can all be uniquely expressed as polynomials

$$(1) \qquad f = a_o + xa_1 + \ldots + x^n a_n \qquad (a_i \in A).$$

As usual we write deg $f = n$ if $a_n \neq 0$ in (1). The additive group of R is just a direct sum of copies of A (by the uniqueness of (1)). To multiply two elements, say $f = \Sigma x^i a_i$, given by (1) and $g = \Sigma x^j b_j$ we have, by distributivity, $fg = \Sigma x^i (a_i x^j) b_j$ and so it will only be necessary to prescribe $a_i x^j$. To ensure that R is again entire, let us assume that

$$(2) \qquad \deg fg = \deg f + \deg g.$$

Then in particular, ax for any $a \in A$ has degree at most 1, so

$$(3) \qquad ax = xa^{\alpha} + a^{\delta},$$

where $a \longmapsto a^{\alpha}$, $a \longmapsto a^{\delta}$ are mappings of A into itself. This is already enough to fix the multiplication in R, for now we can work out ax^r by induction on r:

$$ax^r = (xa^{\alpha} + a^{\delta})x^{r-1} = [x^2 a^{\alpha^2} + x(a^{\alpha\delta} + a^{\delta\alpha}) + a^{\delta^2}]x^{r-2} = \ldots$$

We derive some consequences from (3):

$$(a + b)x = x(a + b)^{\alpha} + (a + b)^{\delta}, \quad ax + bx = xa^{\alpha} + a^{\delta} + xb^{\alpha} +$$

hence

$$(4) \qquad (a + b)^{\alpha} = a^{\alpha} + b^{\alpha}, \qquad (a + b)^{\delta} = a^{\delta} + b^{\delta}, \quad \text{and}$$

$$(ab)x = x(ab)^{\alpha} + (ab)^{\delta}, \; a(bx) = a(xb^{\alpha} + b^{\delta}) = xa^{\alpha}b^{\alpha} + a^{\delta}b^{\alpha} + a$$

so

$$(5) \qquad (ab)^{\alpha} = a^{\alpha}b^{\alpha}, \qquad (ab)^{\delta} = a^{\delta}b^{\alpha} + ab^{\delta}.$$

Further, $1x = x1$, therefore

$$(6) \qquad 1^{\alpha} = 1, \quad 1^{\delta} = 0,$$

and since ax has degree 1 for $a \neq 0$,

$$(7) \qquad a^{\alpha} = 0 \quad \Rightarrow \quad a = 0.$$

From (4)-(7) we see that α is an injective endomorphism of A and δ is an α-*derivation* of A, i.e. a mapping such that

$$(8) \qquad (a + b)^{\delta} = a^{\delta} + b^{\delta}, \qquad (ab)^{\delta} = a^{\delta}b^{\alpha} + ab^{\delta}.$$

We note that (8) entails $1^{\delta} = 0$, by putting $a = b = 1$ in the second equation. Conversely, let A be an entire ring with an injective endomorphism α and an α-derivation δ. Then the set of all expressions (1) can be made into a ring by defining addition componentwise and multiplication by the commutation rule (3). The resulting ring R is again entire (because the degree on it satisfies (2)). It is called the *skew polynomial ring* in x over A (associated with α,δ) and is denoted by

$$A[x;\alpha,\delta].$$

When $\delta = 0$ we also write $A[x;\alpha]$ instead of $A[x;\alpha,0]$; if moreover $\alpha = 1$, we obtain the polynomial ring in a central indeterminate over A, also written $A[x]$.

In matrix notation the commutation rule (3) can be written

$$a(x \quad 1) = (x \quad 1)\begin{pmatrix} a^{\alpha} & 0 \\ a^{\delta} & a \end{pmatrix}$$

and the conditions (4)-(6) may be summed up by saying that the mapping of A into the matrix ring A_2 defined by

$$(9) \qquad a \longmapsto \begin{pmatrix} a^{\alpha} & 0 \\ a^{\delta} & a \end{pmatrix}$$

is a ring homomorphism. More precisely, it is a homomorphism into the ring of lower 2 x 2 triangular matrices over A, which in a suggestive notation may be written $\begin{pmatrix} A & 0 \\ A & A \end{pmatrix}$. Suppose now that A is a right Ore domain with field of fractions K, then any injective endomorphism α of A extends to a unique endomorphism of K, again denoted by α. An α-derivation δ defines a homomorphism (9) which by functoriality extends to a homomorphism $K \longrightarrow \begin{pmatrix} K & 0 \\ K & K \end{pmatrix}$, say $u \longmapsto \begin{pmatrix} u^{\alpha} & 0 \\ u' & u \end{pmatrix}$. Clearly $u \longmapsto u'$ is an α-derivation on K extending δ and we shall write u^{δ} instead of u'. Thus we have shown (using (9)) that any α-derivation of a right Ore domain extends to a unique α-derivation of the field of fractions. This remark will be useful later.

Let A be any ring with an endomorphism α, then for each $c \in A$ the mapping

$$\Delta_c : a \longmapsto ac - ca^{\alpha}$$

is easily seen to be an α-derivation; it is called the *inner* α-derivation induced by c. A derivation which is not inner is called *outer*. The construction of the skew polynomial ring shows that any α-derivation δ on A may be made inner by going over to $A[x;\alpha,\delta]$ (which can of course be defined even if A has zero-divisors and α fails to be injective). In this ring δ is induced by x; moreover, if δ was already inner on A, say $\delta = \Delta_c$ then on writing $y = x - c$ we have $A[x;\alpha,\delta] = A[y;\alpha]$, as is easily checked.

Let K be a field, then any endomorphism of K is injective,

for its kernel is a proper ideal of K. Thus for any endomorphism α and any α-derivation δ, the skew polynomial ring $R = K[x;\alpha,\delta]$ is entire. As in the commutative case we can show that this is a principal right ideal domain. For let $\mathfrak{a}$ be a right ideal; if $\mathfrak{a} \neq 0$, pick a monic polynomial f say, of least degree in $\mathfrak{a}$. Then every $g \in \mathfrak{a}$ can be written as $g = fq + r$, where deg r < deg f. Since $r = g - fq \in \mathfrak{a}$, it follows that $r = 0$ and so $g \in fR$. Thus $\mathfrak{a} = fR$ and we have proved the first part of

Proposition 1.3.1. *Any skew polynomial ring* $K[x;\alpha,\delta]$ *over a field* K *is a principal right ideal domain. It is a principal left ideal domain if and only if* α *is an automorphism.*

To prove the second part, assume first that α is an automorphism. Then if $a^{\alpha} = b$, we have $a = b^{\beta}$, where $\beta = \alpha^{-1}$, and the commutation relation (3) can be rewritten as $b^{\beta}x = xb + b^{\beta\delta}$, i.e.

$$(10) \quad xb = b^{\beta}x - b^{\beta\delta}.$$

Now it follows by symmetry that R is a principal left ideal domain. Conversely, assume that α is not an automorphism, then it is not surjective and so there exists $c \in K$, $c \notin K^{\alpha}$. We assert that $Rx \cap Rxc = 0$, for if not, then for some $f,g \in R^*$,

$$(11) \quad fx = gxc.$$

Comparing degrees we see that deg f = deg g = n say. Let $f = x^n a + \ldots$, $g = x^n b + \ldots$, then by comparing highest terms in (11) we find $a^{\alpha} = b^{\alpha}c$, hence $c = (b^{-1}a)^{\alpha}$, and this contradicts the choice of c. So we have shown that $Rx \cap Rxc = 0$, i.e. R is not a left Ore domain. Now the assertion follows from the fact (proved below) that every left Noetherian domain is a left Ore domain. For if R were left principal, it would be left Noetherian and hence left Ore, but we have just seen that this is not so. ■

We still have to prove the assertion made in the course of the proof about left Noetherian domains. Translating to the right, we find that what we need is

Proposition 1.3.2. *Any right Noetherian domain is right Ore.*

Proof. Let R be a right Noetherian domain and $a,b \in R^*$, then $\Sigma_o^\infty a^i bR$ is finitely generated, hence for some $n \geq 1$,

$$a^n b = bc_o + abc_1 + \ldots + a^{n-1}bc_{n-1}.$$

Since $a^n b \neq 0$, not all the c_i vanish; let c_k be the first non-zero coefficient, then we can cancel a^k and obtain

$$a^{n-k}b = bc_k + abc_{k+1} + \ldots + a^{n-k-1}bc_{n-1},$$

so $bc_k \in aR$, i.e. $aR \cap bR \neq 0$, and this is just the right Ore condition. ■

This proof (which goes back to Goldie) actually shows a little more: In an entire ring R, if $aR \cap bR = 0$ for $a,b \neq 0$, then the elements $a^n b$ $(n = 0,1,\ldots)$ are right linearly independent over R. For if $\Sigma a^i bc_i = 0$, and c_k is the first non-zero coefficient, then as before we find that $bc_k \in aR$. Hence if $aR \cap bR = 0$, we can find a free right ideal of countable rank in R. This proves the

Corollary. An integral domain is either a right Ore domain or it contains free right ideals of countable rank. ■

Prop. 1.3.1 shows that any skew polynomial ring $K[x;\alpha,\delta]$ over a field is right Noetherian, and hence right Ore by Prop. 1.3.2. Therefore it has a unique field of fractions, which we shall denote by $K(x;\alpha,\delta)$.

There is an interesting application of the last Cor., due to Jategaonkar [69'] and independently to Koševoi [70].

Proposition 1.3.3. *Let* R *be an entire ring with centre* C. *Then* R *is either a left and right Ore domain or it contains a free* C*-algebra of countable rank.*

Proof. Suppose that R is not right Ore; it will be enough to find two elements x,y which are free, for then the elements $x^i y$ form an infinite free generating set. We choose $x,y \in R^*$ such that $xR \cap yR = 0$ and claim that the C-algebra generated by x and y is free. If not, let f = 0 be a non-trivial relation of least degree between x and y. This has the form

$$(12) \quad \alpha + xa + yb = 0 \qquad a,b \in R,\ \alpha \in C.$$

Here a,b are not both zero, since they have lower degree than f. Say, $b \neq 0$, then on multiplying (12) on the right by x we have $\alpha x + xax + ybx = 0$, i.e. $ybx = x(-\alpha - ax) \in xR \cap yR$, and $ybx \neq 0$, a contradiction, which shows x,y to be free over C. ∎

The result can be used to embed a free algebra in a field (as both Jategaonkar and Koševoi have observed). For let K be a field with a non-surjective endomorphism α (e.g. the rational function field k(t) with endomorphism $f(t) \longmapsto f(t^2)$) and form the skew polynomial ring $R = K[x;\alpha]$. This is an entire ring; by Prop. 1.3.1 it is not left Ore and so it contains a free C-algebra on two free generators, where C is the centre of R. But R is right Ore and so it can be embedded in a field; this then provides an embedding of the free algebra (of countable rank) in a field.

In spite (or perhaps because) of its simplicity this construction is of limited use, because not every automorphism of $C\langle x,y\rangle$ can be extended to an automorphism of the field of fractions, constructed here. One application of this construction, due to J.L. Fisher [71] is to show that $C\langle x,y\rangle$ has many different fields of fractions. Let $A = k[t]$ be the usual (commutative) polynomial ring with endomorphism α_n: $f(t) \longmapsto f(t^n)$, and consider the subring of $R = A[x;\alpha_n]$

generated by x and $y = xt$ over k, for $n > 1$. Since t is not in the image of $k(t)$ under α_n, it follows as in the proof of Prop. 1.3.1 that $Rx \cap Ry = 0$, hence the subalgebra on x and y over k is free, and for different n we clearly get distinct (i.e. non-isomorphic) embeddings, because

$$x^{-1}yx = tx = xt^n = x(x^{-1}y)^n = (yx^{-1})^n x, \text{ so } x^{-1}y = (yx^{-1})^n.$$

2·Topological methods

2.1 Power series rings

In the commutative case the familiar power series ring $k[[x]]$ may be regarded as the completion of the polynomial ring with respect to the "x-adic topology", i.e. the topology obtained from the powers of the ideal generated by x. It is no problem to extend this concept to the ring $k[x;\alpha]$, but when there is a non-zero derivation we face a difficulty. The above topology may be described in terms of the order-function.

$$o(f) = r \quad \text{if} \quad f = x^r a_r + x^{r+1} a_{r+1} + \dots + x^n a_n \qquad (a_r \neq 0).$$

Now it turns out that when $\delta \neq 0$, multiplication is not continuous in the x-adic topology, as the formula

$$(1) \qquad a.x = xa^{\alpha} + a^{\delta}$$

shows, and any attempt to construct the completion directly will fail. One way out of this difficulty is to introduce $y = x^{-1}$ and rewrite (1) as a commutation formula for y. We then get

$$(2) \qquad ya = a^{\alpha}y + ya^{\delta}y.$$

Owing to the inversion we now have to shift coefficients to the left (unless of course α happens to be an automorphism) and as (2) shows we cannot usually do this completely in the polynomial ring, but we can do it to any desired degree of accuracy by applying (2) repeatedly:

$$ya = a^{\alpha}y + a^{\delta\alpha}y^2 + ya^{\delta^2}y^2$$

$$(3) \qquad = a^{\alpha}y + a^{\delta\alpha}y^2 + a^{\delta^2\alpha}y^3 + \ldots + a^{\delta^{n-1}\alpha}y^n + ya^{\delta^n}y^n.$$

If δ is locally nilpotent, i.e. for each $a \in k$ there exists n such that $a^{\delta^n} = 0$, this can be used as a commutation formula in the skew polynomial ring. This kind of formula has been studied by T.H.M. Smits [68] (cf. also Cohn [71"],Ch.0). But in any case, in the power series ring we can pass to the limit in (3) and obtain the formula

$$(4) \quad ya = a^{\alpha}y + a^{\delta\alpha}y^2 + a^{\delta^2\alpha}y^3 + \ldots$$

The ring obtained in this way is clearly an integral domain and the set consisting of powers of y is a left denominator set, by (4), so we can form the ring of fractions, which is in effect the ring of (skew) formal Laurent series in y.

If K is a field with a surjective derivation δ, then $K(x;1,\delta)$ is a field with a surjective inner derivation, and we can also obtain a field with this property by taking skew Laurent series. Thus if the commutation formula is $cx^{-1} = x^{-1}c - c^{\delta}$, then on writing $[a,b] = ab - ba$, we have $[x^{-1},c] = c^{\delta}$ and hence

$$[x^{-1},\Sigma x^i c_i] = \Sigma x^i [x^{-1},c_i] = \Sigma x^i c_i^{\delta}.$$

Since δ is surjective, every element has this form and the assertion follows. To get a surjective derivation we need only take a function field in which every function is a derivative, or more algebraically, a differentially closed field (cf.e.g. Sacks [72], Shelah [72]). This answers a question first raised by Kaplansky [70]; he asked for a field in which every element is a sum of commutators and B. Harris [58] answered this by constructing a field in which every element is a commutator. Such a field is the union of the ranges of its inner derivations, and it can also be constructed very simply as follows (Bokut' [63]): Given any field K, the rational function field $L = K(x)$ in a central indeterminate x admits the derivation $f \longmapsto f'$ (the usual derivative) and in the skew function field $K_1 = L(y;1,')$ we have $[x,y] = 1$, hence $[ax,y] = a$ for any $a \in K$. If we repeat this process we obtain a field $K_2 \supset K_1$ and every element of K_1 is a commutator in K_2. Thus we can form an ascending chain $K_1 \subset K_2 \subset \ldots$ whose union is a field which is the union of the ranges of its inner derivations.

In finite characteristic Lazerson [61] has constructed a field with a surjective inner derivation: Given any field k, of characteristic $p \neq 0$, adjoin commuting indeterminates $x_1, x_2, \ldots$ to form $K = k(x_1, x_2, ..)$ with derivation δ such that $x_i^\delta = x_{i-1}$ $(i > 1)$, $x_1^\delta = 1$. Put $L = K(t;1,\delta)$ then δ is induced by t, hence δ^q is an inner derivation induced by t^q, for any $q = p^n$, and it annihilates anything involving only $x_1, \ldots, x_{q-1}$. Thus, given $a \in L$, there exists $q = p^n$ such that $[a,t^q] = 0$ and so $[ax_q, t^q] = a$. This shows that δ is surjective. By taking ultra-products of such fields for varying p we obtain fields of characteristic 0 with surjective derivations. For other constructions see Cohn [73'''].

There is an important generalization of the power series method, to which we now turn. Let G be a group and consider the group algebra kG over a commutative field k. When is kG embeddable in a field? Clearly a necessary condition is that it should be entire, and for this it is necessary for G to be torsion free. For if $u \in G$ is of order n, then

$$(u - 1)(u^{n-1} + u^{n-2} + \ldots + u + 1) = 0.$$

In the abelian case this condition on G is also sufficient. For if G is torsion free abelian, it can be totally ordered (regard G as $\mathbb{Z}$-space, embed it in a $\mathbb{Q}$-space and use a lexicographic ordering with respect to an ordered basis). When G is totally ordered, kG is clearly entire: Let $a = a_1s_1 + \ldots + a_ms_m$ with $a_i \in k$, $s_i \in G$ and $s_1 < \ldots < s_m$, similarly $b = b_1t_1 + \ldots + b_nt_n$ ($b_j \in k$, $t_j \in G$, $t_1 < \ldots < t_n$), then $ab = a_1b_1s_1t_1 + \ldots$, where the dots represent terms greater than s_1t_1, hence $ab \neq 0$. Thus we have

Proposition 2.1.1. *Let G be an abelian group, then the group algebra* kG *(over any commutative field* k*) is embeddable in a field if and only if* G *is torsion free.* ■

In the non-commutative case little is known; it is not even known whether kG is entire for any torsion free G. But Farkas and Snider [76] have recently proved that this is the case when G is polycyclic (i.e. soluble with maximum condition on subgroups); since kG is Noetherian in this case, it is then embeddable in a field. In another direction J. Lewin and T. Lewin [a] have shown (using methods of Magnus and some results from Ch.4 below) that for any torsion free group G with a single defining relation the group algebra kG can be embedded in a field.

It has long been known that Prop.2.1.1 can be generalized to non-abelian groups which are ordered. In that case we can form a kind of power series ring k((G)) which turns out to be a field. This was first proved by Hahn [07] for abelian groups with an archimedean ordering and then generally by Mal'cev [48] and independently, Neumann [49']. The result was put in a general algebraic setting by Higman [52]; his proof (with some simplifications) is given in Cohn [65]. Let us briefly describe the construction without entering into the details of the proof.

Consider the k-space k^G of all k-valued functions on G; it contains the group algebra as subspace, in fact $a = (a_g)$

belongs to the group algebra precisely if its *support*

$$D(a) = \{g \in G \mid a_g \neq 0\}$$

is finite. Now the multiplication on kG cannot in any natural way be extended to k^G; if $a = (a_g)$, $b = (b_g)$, then we should have

$$(5) \qquad (ab)_g = \sum_{h \in G} a_h b_{h^{-1}g},$$

and there is no guarantee that the sum on the right is finite. Let k((G)) be the subset of k^G consisting of all elements with well-ordered support (in the ordering on G). If $a,b \in k((G))$, then the sum on the right of (5) is finite for each g, for the h such that $a_h \neq 0$ form an ascending chain, so the $h^{-1}g$ (g fixed) form a descending chain and hence only finitely many such $b_{h^{-1}g}$ are non-zero. Moreover, it is easily seen that $ab \in k((G))$, so that the latter forms in fact a ring, with kG as subring. The theorem of Mal'cev and Neumann asserts that k((G)) is actually a field. Thus each element of k((G)) has the form $\Sigma a_u u$ and if $u_o \in G$ is the least element for which $a_{u_o} \neq 0$, then we can write $f = a_{u_o} u_o(1 - g)$, where g has support consisting only of elements > 1. Now what has to be proved is that $1 + g + g^2 + \ldots \in k((G))$; once that is established we clearly have

$$f^{-1} = (1 + g + g^2 + \ldots)\, u_o^{-1} a_{u_o}^{-1} .$$

In order to apply this result to embed free algebras in fields we use the fact that the free group can be totally ordered (cf. e.g. Fuchs [63]). Briefly the proof goes as follows:

Let F be the free group on a set X, taken to be finite

for simplicity, and let $b_1, b_2, \ldots$ be the sequence of basic commutators in X. Denote by $\gamma_t(F)$ the tth term of the lower central series of F and let $b_1, b_2, \ldots, b_w$ be all the basic commutators of weight $< t$, then as is well known, every element $a \in F$ has a unique expression

$$a \equiv b_1^{\alpha_1} b_2^{\alpha_2} \ldots b_w^{\alpha_w}$$

$$(\bmod\ \gamma_t(F)) \qquad\qquad (\alpha_i \in \mathbf{Z})$$

(cf. e.g. M. Hall [59]), and we can therefore represent each $a \in F$ by an infinite product

$$a = b_1^{\alpha_1} b_2^{\alpha_2} \ldots$$

Now write $a > 1$ whenever the first non-zero of $\alpha_1, \alpha_2, \ldots$ is positive; this provides a total ordering of F. The same method works for free groups of infinite rank (taking the free generating set to be well-ordered).

It is clear that $k\langle X\rangle$, the free k-algebra on X, can be embedded in kF, the group algebra of the free group on X, and since kF is embedded in the 'power series field' k((F)) just constructed, we have another embedding of $k\langle X\rangle$ in a field. If instead of the free group F we take the free metabelian group G, i.e. the group defined by the law $((u,v),(w,x)) = 1$ (where $(x,y) = x^{-1}y^{-1}xy$), we can still embed $k\langle X\rangle$ in kG, at least when card (X) = 2, the crucial case (Moufang [37]). Moreover, G can again be ordered, so we have another embedding of $k\langle X\rangle$ in a field, and these two embeddings of $k\langle X\rangle$ in k((F)) and k((G)) are clearly distinct.

Of course there are other simpler ways of finding non-isomorphic fields of fractions of $k\langle X\rangle$, e.g. that by Fisher described earlier. We note that in the above construction the free metabelian group cannot be replaced by a free nilpotent group of any class, for the free semigroup on X cannot be embedded in any nilpotent group on X, by results of Mal'cev (cf. Lyapin [60]).

As an application of the Mal'cev-Neumann construction we derive the one-sided principal ideal domains first obtained by Jategaonkar. We have seen that a polynomial ring over a field is a principal ideal domain, and it is clear that the condition is necessary, i.e. if a polynomial ring is principal, the coefficient ring must be a field. This is true even for skew polynomial rings relative to an automorphism, but for a non-surjective endomorphism it need not hold. The precise conditions were determined by Jategaonkar [69]:

Theorem 2.1.2. *Let* A *be a ring with an endomorphism* S *and put* $R = A[x;S]$. *Then* R *is a principal right ideal domain if and only if* A *is a principal right ideal domain and* S *maps* A* *into* U(A), *the group of units of* A.

Proof. If R is principal, so is A because it is a retract of R, i.e. a subring which is also a homomorphic image (put $x = 0$). Further, for any $a \in A^*$ we have $aR + xR = cR$, where c is the highest common left factor of a and x. It follows that c has degree 0 (as factor of a), so $x = cf$, where f has degree 1, say $f = xd + e$. Now $x = cxd + ce = xc^S d + ce$, which shows that $ce = 0$, $c^S d = 1$, so c^S is a unit. Let $au + xv = c$, then putting $x = 0$ we see that a is associated to c, hence a^S is associated to c^S, a unit, so a^S is a unit, as claimed.

Conversely, if the given conditions hold, R is clearly entire. Let $\mathfrak{a}$ be a right ideal in R. When $\mathfrak{a} = 0$, there is nothing to prove; otherwise let n be the least degree of polynomials occurring in $\mathfrak{a}$. The leading coefficients of polynomials of degree n in $\mathfrak{a}$ form with 0 a right ideal in A, generated by a say. Let $f = x^n a + \ldots \in \mathfrak{a}$, then $a^S \in U(A)$ and hence $fx = x^{n+1} a^S + \ldots$ has a unit as highest coefficient. It follows that $\mathfrak{a}$ contains a monic polynomial of degree n+1 and so also of all higher degrees. Now it is clear that

$\mathfrak{a} = fR$, hence R is a principal right ideal domain. ■

This result shows that under favourable circumstances one may iterate the polynomial ring construction and still get a principal right ideal domain, and this suggests the following definitions. By a *J-skew polynomial ring* one understands a skew polynomial ring $A[x;S]$ such that S is injective and satisfies Jategaonkar's condition:

$$A^S \subseteq U(A) \cup \{0\}.$$

E.g. this condition holds whenever A is a field; what is of interest is that there are other cases. It is easily seen that any J-skew polynomial ring over A is entire if A is.

Now let R be a ring and τ an ordinal number, then R is called a *J-ring of type* τ if R has a chain of subrings R_α $(\alpha \leq \tau)$ such that

(i) $R_o = U(R) \cup \{0\}$ (hence R_o is a field),

(ii) $R_{\alpha+1}$ is a J-skew polynomial ring over R_α for all $\alpha<\tau$,

(iii) $R_\alpha = \bigcup_{\beta<\alpha} R_\beta$ for any limit ordinal $\alpha\leq\tau$,

(iv) $R = R_\tau$.

Explicitly we have $R_{\alpha+1} = R_\alpha [x_\alpha;S_\alpha]$ and it follows from the definition that each element c of R can be uniquely written as

$$(6) \qquad c = \Sigma x_{\alpha_1} \ldots x_{\alpha_r} c_{\alpha_1 \ldots \alpha_r} \quad (c_{\alpha_1 \ldots \alpha_r} \in R_o, \alpha_1 \geq \ldots \geq \alpha_r).$$

It is easily verified by induction that $U(R_\alpha) = U(R_o)$, hence we have the

Corollary. Any J-ring (of any type τ*) is a principal right ideal domain.* ■

It turns out that J-rings can be characterized as integra

domains with Euclidean algorithm (generally transfinite) and unique remainder (cf. Lenstra [74]). We shall see that for any τ there are J-rings of type τ. Such rings form a useful source of counter-examples; they were constructed by Jategaonkar to provide examples of (i) a principal right ideal domain in which there are non-units with arbitrarily long factorizations (only rather special examples, in effect J-rings of type 2, were known earlier, cf. Cohn [67]), (ii) a ring with left and right global dimensions differing by an arbitrary integer (the largest known difference had been 2 before, cf. Small [65]), (iii) a left but not right primitive ring (such a ring was first constructed by Bergman [64], answering a question of Jacobson [56], but Jategaonkar's example is more direct. See also Brungs [69] for some remarkable properties of this construction).

Skew polynomial rings over a field are J-rings of type 1; J-rings of type 2 can be obtained by an ad hoc construction (Cohn [67]) but beyond this the general case is no harder than the finite case. Moreover, one cannot use induction directly, since the coefficient ring depends essentially on the order type. Jategaonkar uses an ingenious argument involving ordinals; below is a direct proof based on the Mal'cev-Neumann construction (cf. Cohn [71"]).

We observe that to achieve the form (6) we need a commutation rule of the form

$$x_\beta x_\alpha = x_\alpha u_{\beta\alpha} \qquad (\beta < \alpha),$$

where $u_{\beta\alpha}$ has to be a unit in R_o. More generally, this must be true for products of x's, which we may as well take in normal form, as in (6). Thus we shall need a formula of the form

$$(7) \quad x_\beta x_{\alpha_1} \ldots x_{\alpha_r} = x_{\alpha_1} \ldots x_{\alpha_r} u_{\beta\alpha_1\ldots\alpha_r} \quad (\alpha_1 \geq \ldots \geq \alpha_r, \alpha_1 > \beta, r \geq 1).$$

It turns out that this is enough to give the required construction. Thus let $X = \{x_\alpha\}$ $(\alpha < \tau)$ be a family of indeterminates, denote by F_X the free group on X and put $E = k((F_X))$. Let K be the subfield of E generated by the elements

$$(8) \quad u_{\beta\alpha_1\ldots\alpha_r} = (x_{\alpha_1}\ldots x_{\alpha_r})^{-1} x_\beta x_{\alpha_1}\ldots x_{\alpha_r} \quad (\alpha_1 \geq \ldots \geq \alpha_r, \alpha_1 > \beta, r$$

as suggested by (7), then we have the following

Lemma 2.1.3. *The centralizer of any* x_γ *in* K *is* k.

Proof. Let G be the subgroup of $F = F_X$ generated by the right hand sides of (8). Each generator has odd length, so that we can speak of the middle factor. When we form a group element of G, the middle factor of any generator cannot be affected by cancellation, hence any element of G begins with a letter x_λ^{-1} and ends with a letter x_μ even after cancellation. It follows that $x_\gamma^n \notin G$ for $n \neq 0$. Now any $a \in K$ has the form $\Sigma a_u u$, where $u \in G$ and conjugation by x_γ maps K into itself:

$$x_\gamma^{-1} u_{\beta\alpha_1\ldots\alpha_r} x_\gamma = u_{\alpha_r\gamma}^{-1}\ldots u_{\alpha_i\gamma}^{-1} u_{\beta\alpha_1\ldots\alpha_{i-1}\gamma} u_{\alpha_i\gamma}\ldots u_{\alpha_r\gamma},$$

if $\alpha_{i-1} \geq \gamma > \alpha_i$. Now x_γ commutes only with x_γ^n in F, so conjugation by x_γ fixes 1 and moves all other elements of G in infinite orbits. Each of these orbits is generated from a single element of G by conjugating by positive powers of x_γ. Hence $x_\gamma^{-1} a x_\gamma = a$ can hold only if $a_u = 0$ for all $u \neq 1$, thus $a = a_1 \in k$ as claimed. ■

Let S_γ be the endomorphism of G induced by conjugation with x_γ. Conjugation is order-preserving (for any ordering of F or G) so this extends to an endomorphism of K, again denoted by S_γ. Thus for any $a \in K$

$$(9) \quad a x_\gamma = x_\gamma a^{S_\gamma}.$$

Let R be the subring of $E = k((F))$ generated by K and all

$x_\alpha (\alpha < \tau)$. By (9) each element of R can be written in the form

(10) $\Sigma x_{\alpha_1} \cdots x_{\alpha_r} a_{\alpha_1 \cdots \alpha_r}$, where $a_{\alpha_1 \cdots \alpha_r} \varepsilon K$.

If the α_i are not already in descending order, then for some $i = 1,2,\ldots,r-1$, we have $\alpha_i < \alpha_{i+1} \geq \alpha_{i+2} \geq \cdots \geq \alpha_r$. Now we can pull x_{α_i} through to the right using (7). By repeating the process, if necessary, we ensure that $\alpha_1 \geq \cdots \geq \alpha_r$ in each term of (10). We claim that with this proviso, the expression (10) is unique. To establish this claim, suppose we have a relation

(11) $\Sigma x_{\alpha_1} \cdots x_{\alpha_r} a_{\alpha_1 \cdots \alpha_r} = 0$, where $a_{\alpha_1 \cdots \alpha_r} \varepsilon K$, $\alpha_1 \geq \cdots \geq \alpha_r$.

Let α be the highest suffix occurring in (11), then we can write (11) as $\Sigma x_\alpha^i c_i = 0$, where each c_i is a polynomial in the x_β $(\beta < \alpha)$. Conjugating by x_α we obtain coefficients $c_i^{S_\alpha}$ which lie in K; thus x_α satisfies an equation over K. We write down the minimal equation:

(12) $x_\alpha^n + x_\alpha^{n-1} b_1 + \cdots + b_n = 0$, where $b_i \varepsilon K$.

If we conjugate by x_α we obtain another monic equation of degree n for x_α and by the uniqueness of the minimal equation it must coincide with (12). Thus $b_i^{S_\alpha} = b_i$ $(i = 1,\ldots,n)$, and by the lemma, $b_i \varepsilon k$, i.e. x_α is algebraic over k. But this is clearly false, and this contradiction shows that all the coefficients in (11) must vanish. Thus (10) is unique and we have constructed a J-ring of type τ.

Of the properties (i)-(iii) mentioned on p.25, (i) is immediate: $x_2 = x_1^n x_2 u_{12}^{-n}$ for any $n \geq 0$. For (ii),(iii) see Jategaonkar [69], for (iii) see also Cohn [71''].

2.2 Inverse limits of Ore domains

We shall only make a brief mention of the next method, since we shall not use it in what follows. Let R be a filtered ring, i.e. a ring with a series of subgroups

$$(1) \qquad \ldots \subseteq R_{-n} \subseteq R_{-n+1} \subseteq \ldots \subseteq R_o \subseteq R_1 \subseteq \ldots \subseteq R_n \subseteq \ldots$$

such that $\cap R_n = 0$, $\cup R_n = R$ and $R_i R_j \subseteq R_{i+j}$. E.g. a polynomial ring is of this form, if R_n denotes the set of polynomials of degree at most n (even skew polynomials). With R we can associate another ring gr R = $\oplus$ $gr_n R$, where $gr_n R = R_n/R_{n-1}$ and multiplication is defined in the natural way: if $\alpha \in R_i$, $\beta \in R_j$ then $\alpha\beta \in R_{i+j}$ and $\alpha\beta$ is determined mod R_{i+j-1} by the residue classes of α (mod R_{i-1}) and β (mod R_{j-1}). This ring gr R is the *associated graded ring*; it may be looked upon as the ring of "leading terms". The next result gives conditions for a filtered ring to be embeddable in a field:

Theorem 2.A. *Let* R *be a filtered ring; if the associated graded ring is a right (or left) Ore domain then* R *is embeddable in a field.*

The proof proceeds by embedding R* in a group, constructed as an inverse limit, and then extending addition to it (cf. Cohn [61]). In this construction it is essential to have a **Z**-filtration (i.e. the terms in (1) are indexed by **Z**), but more general sets of conditions have been examined by Dauns [70].

A first application is again to embed the free algebra k<X> in a field. Let L be the free Lie algebra on X over k, then its universal associative envelope U is well known to be isomorphic to k<X>. Moreover, U is filtered by the powers of L: $0 \subset L \subset L^2 \subset \ldots$ and the associated graded ring is just the polynomial ring in a basis of L over k. Clearly this is an Ore domain (even commutative), hence

Th.2.A above can be applied to embed $U = k\langle X\rangle$ in a field. Like the power series method this produces a topological field (the topology being defined by the inverse limit construction) and whether a given automorphism of $k\langle X\rangle$ can be extended to the field of fractions depends on whether it is continuous.

Another less obvious application is to the construction of skew field extensions of different left and right dimensions (Cohn [61']), but we shall achieve the same end more simply by using method 4 in 5.5 below.

3· Extensions of finite degree

3.1 Generalities

It is a trivial observation that if k is a commutative field and α is algebraic over k, then the field $k(\alpha)$ generated by α over k has finite degree. Moreover, any finite number of algebraic elements over k generate an extension of finite degree. For skew fields there is no corresponding statement; here the extensions of finite degree are very much more complicated and there is no simple way of producing them all, as in the commutative case. In fact there could be ambiguity about what is meant by an extension of finite degree.

Let K be a (skew) field and E a subfield, then K may be regarded as left or right vector space over E and correspondingly we have two numbers (possibly infinite), the *left* and *right degree*:

$$[K:E]_L \quad \text{and} \quad [K:E]_R.$$

Later we shall meet examples where one of these numbers is finite and the other infinite, but no examples are known where both are finite and different. We shall call $[K:E]_R$ the *degree* of K over E and call the extension K/E *finite* when its degree is finite. As in the commutative case one has the Transitivity formula. If $F \subseteq E \subseteq K$ are any fields, then

(1) $$[K:F]_R = [K:E]_R[E:F]_R,$$

whenever either side is finite.

The proof is as in the commutative case, by showing that if $\{u_\lambda\}$ is a right E-basis for K and $\{v_i\}$ a right F-basis for E, then $\{u_\lambda v_i\}$ is a right F-basis for K. ■

At least one of our difficulties disappears for extensions of finite degree, the difference between zero-divisors and non-units:

Proposition 3.1.1. *Let* K *be a field and* A *a* K*-ring of finite (right) degree over* K*, then any left non-zerodivisor of* A *is a unit, hence if* A *is entire, it is a field.*

Proof. Let $a \in A$ and suppose that $ax = 0 \Rightarrow x = 0$. Then the mapping $\lambda_a : x \longmapsto ax$ is injective, and it is clearly right K-linear, on a finite-dimensional K-space, hence it is surjective, and so $ab = 1$ for some $b \in A$. Now b is again a left non-zerodivisor: if $bx = 0$, then $x = abx = 0$. Hence there exists $c \in A$ such that $bc = 1$, and so $c = abc = a$ and this shows that $ab = ba = 1$, i.e. a is a unit. The rest is clear. ■

There is one important case where left and right degrees are the same.

Theorem 3.1.2. *Let* K *be a field of finite degree over its centre. Then the left and right degrees over any subfield coincide.*

Proof. Let E be any subfield of K and denote the centre of K by C. By hypothesis K is a C-algebra of finite degree, and it is clear that $A = EC = \{\Sigma x_i y_i \mid x_i \in E, y_i \in C\}$ is a subalgebra. If we regard A as E-ring, we can choose a basis of A as left E-space consisting of elements of C; then this will also be a right E-basis for A, hence

$$(2) \qquad [A:E]_L = [A:E]_R.$$

Now A is a C-algebra of finite degree, entire as subalgebra of K, hence A is a field. By (1)

$$[K:C] = [K:A]_L[A:C] = [K:A]_R[A:C].$$

Since $[K:C]$ is finite, so is $[A:C]$. If we divide by $[A:C]$ and multiply by (2) we get (on using (1) again)

$$[K:E]_L = [K:E]_R. \blacksquare$$

3.2 The Sweedler predual and the Jacobson-Bourbaki correspondence

The aim of this section is to establish the Jacobson-Bourbaki correspondence theorem, used later for Galois theory and also useful elsewhere. We shall first prove the Sweedler correspondence theorem on corings, see Sweedler [75].

We begin by explaining the notion of a coring. Let A be a ring. By an A-*coring* we understand an A-bimodule M together with A-bimodule maps $\Delta:M \longrightarrow M \otimes_A M$, $\varepsilon:M \longrightarrow A$, such that the following diagrams commute:

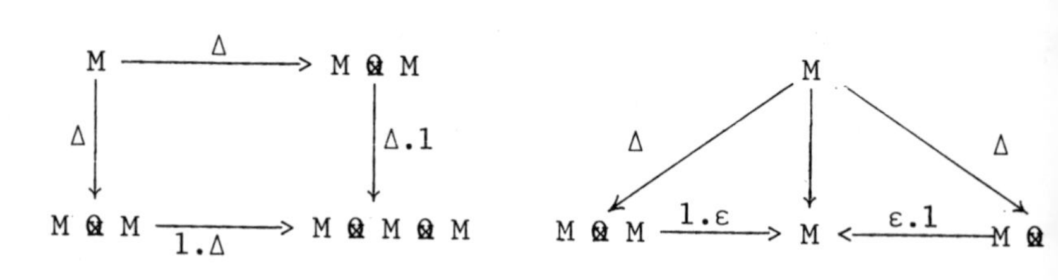

Example. Let $\Phi:A \longrightarrow B$ be a ring homomorphism (i.e. B is an A-ring), put $C = B \otimes_A B$, and define $\varepsilon:C \longrightarrow B$ by $u \otimes v \longmapsto uv$, $\Delta:C \longrightarrow C \otimes_B C$ by $u \otimes v \longmapsto u \otimes 1 \otimes v$ (observe that $C \otimes_B C = B \otimes_A B \otimes_B B \otimes_A B \cong B \otimes_A B \otimes_A B$). Then we have a B-coring structure on C, as is easily checked.

Given any A-ring B, the coring $C = B \otimes_A B$ just constructed is called the *standard B-coring* over A. In particular, for $B = A$, $\Phi = 1$ we have $C = B$; this is the *trivial* B-coring over B.

A *coring map* between A-corings is a mapping compatible with Δ, ε. E.g. if M is any A-coring, $\varepsilon : M \longrightarrow A$ is a coring map to the trivial coring. A *coideal* of M is an A-subbimodule J of M such that $\varepsilon(J) = 0$, $\Delta(J) \subseteq J \otimes M + M \otimes J$. This allows a coring structure to be defined on M/J, and it is not hard to check that the kernel of any coring map is a coideal (one needs the fact that for modules U, V with submodules U', V' and natural homomorphisms $\alpha : U \longrightarrow U/U'$, $\beta : V \longrightarrow V/V'$, the kernel of $\alpha \otimes \beta$ is $\mathrm{im}(U' \otimes V) + \mathrm{im}(U \otimes V')$).

An element g in a coring M is called *grouplike* if $\varepsilon(g) = 1$, $\Delta(g) = g \otimes g$. E.g. the standard B-coring $B \otimes_A B$ has the *standard grouplike* $1 \otimes 1$.

Proposition 3.2.1. *Any coring map takes grouplikes to grouplikes. Given a homomorphism* $\Phi : A \longrightarrow B$, *let* $C = B \otimes_A B$ *be the standard* B-*coring over* A, *then for any* B-*coring* P *there is a natural bijection (of sets)*

$$\mathrm{Hom}(C,P) \cong G_A(P),$$

where Hom *denotes the set of* B-*coring maps and*

$$G_A(P) = \{g \in P \mid g \textit{ grouplike and } ga = ag \textit{ for all } a \in A\}.$$

Proof. The first part is clear, to prove the second we note that under any coring map $C \longrightarrow P$, $1 \otimes 1$ maps to a grouplike centralizing A. Conversely, given $g \in G_A(P)$, the rule $\Sigma a_i \otimes b_i \longmapsto \Sigma a_i g b_i$ defines a map $C \longrightarrow P$ which is easily seen to be a coring map. ■

We shall often write $g_{B/A}$ for $1 \otimes 1$ in $B \otimes_A B$.

It is a natural question to ask if every B-coring is standard over some subring. Since $B \otimes_A B$ is generated as B-bimodule by the grouplike $g_{B/A}$, we shall limit ourselves to corings generated by a single grouplike. Then the answer is 'yes', provided that B is a skew field:

Proposition 3.2.2. *Let* K *be a skew field and* M *any* K*-coring. Given a grouplike* $g \in M$*, write* $D = \{x \in K \mid xg = gx\}$*, then* D *is a subfield of* K *and the standard coring map (Prop.* 3.2.1)

$$(1) \qquad \zeta : K \otimes_D K \longrightarrow M \qquad\qquad 1 \otimes 1 \longmapsto g$$

is injective; it is an isomorphism if $M = KgK$.

Proof. Clearly D is a field. Suppose ζ is not injective, then there exists $x = \Sigma a_i \otimes b_i \neq 0$ such that

$$(2) \qquad \sum_1^n a_i g b_i = 0 \qquad\qquad (a_i, b_i \in K),$$

and we assume that n, the number of terms in (2), is minimal. Moreover, $0 = \varepsilon(\Sigma a_i g b_i) = \Sigma a_i b_i$, hence $n > 1$ and after multiplying by a_1^{-1} we may take $a_1 = 1$. Now by minimality $a_1 = 1$, $a_2, \ldots, a_n$ are right D-independent and $a_2 g \neq g a_2$, hence there is a right K-space map $\Theta : M \longrightarrow K$ such that $\Theta(a_2 g - g a_2) \neq 0$.

Consider $x' = \Sigma_1^n \Theta(a_i g - g a_i) \otimes b_i$. By the independence of the b_i we have $x' \neq 0$, but $a_1 g - g a_1 = 0$ and so x' has fewer than n terms. Now let us chase x around the diagram

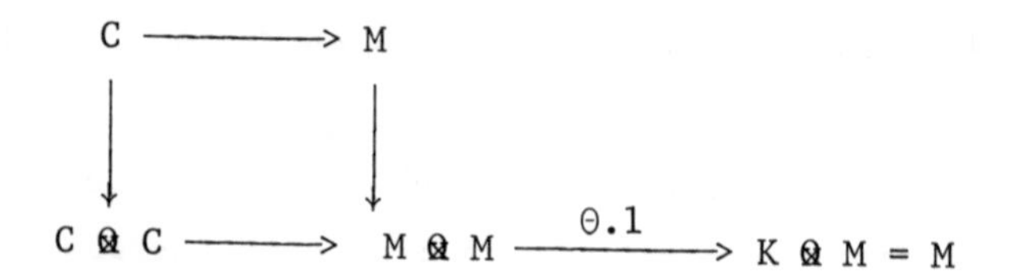

where $C = K \otimes_D K$. Going across and down we get 0, going down and across we get $\Sigma\Theta(a_i g) g b_i$, hence

$$(3) \qquad \Sigma\Theta(a_i g) g b_i = 0.$$

Now

$$\zeta(x') = \Sigma\Theta(a_i g - ga_i)\, gb_i = \Sigma\Theta(a_i g)gb_i - \Sigma\Theta(g)a_i gb_i.$$

Here the first sum vanishes by (3) and the second by (2). Thus $\zeta(x') = 0$, but $x' \neq 0$ and x' has $< n$ terms, a contradiction. Therefore (1) is injective; clearly it is surjective, and hence an isomorphism, whenever $M = KgK$. ■

Taking the case $M = KgK$, we get the

Corollary. Every K*-coring over a field* K*, generated by a grouplike is standard.*

We can now prove the first correspondence theorem:

Theorem 3.2.3 *(Sweedler correspondence theorem). Given skew fields* $F \subseteq K$*, let* $M = K \otimes_F K$ *be the standard* K*-coring over* F*, write* $\mathcal{C}$ *for the set of coideals in* M *and* $\mathcal{D}$ *for the set of fields* D *such that* $F \subseteq D \subseteq K$*, then there is an order-preserving bijection* $\mathcal{C} \longleftrightarrow \mathcal{D}$ *defined by*

$$\alpha: D \longmapsto D^+ = \ker(M = K \otimes_F K \longrightarrow K \otimes_D K),$$
$$\beta: J \longmapsto J^+ = \{x \in K \mid x.\ \pi(g_{K/F}) = \pi(g_{K/F}).x\},$$

where $\pi: M \longrightarrow M/J$.

Proof. As kernel of a coring map D^+ is a coideal. If L is any K-bimodule and $c \in L$, then $\{x \in K \mid xc = cx\}$ is clearly a subfield of K, so J^+ is a subfield of K, and it contains F.

$\alpha\beta = 1$. Given $D \in \mathcal{D}$, $D^+ = J_D$ is the kernel of the natural map $K \otimes_F K \longrightarrow K \otimes_D K$ and $g_{K/F} \longmapsto g_{K/D}$. Now $x \in D^{++} \Longleftrightarrow x \otimes 1 = 1 \otimes x \Longleftrightarrow x \in D$.

$\beta\alpha = 1$. Given $J \in \mathcal{C}$, put $D = J^+$ and let $\pi: M \longrightarrow M/J$ be the natural map, then $c = \pi(g_{K/F})$ centralizes D, by definition of D. Moreover, c generates M/J, hence $M/J \cong K \otimes_D K$ by Prop. 3.2.2, where the homomorphisms of M correspond in this isomorphism, so $J = D^+ = J^{++}$. ■

In this correspondence F was any field, e.g. we can take

it to be the prime subfield P of K, then we obtain a bijection between the coideals of $K \otimes_P K$ and all subfields of K.

To obtain a case of Jacobson-Bourbaki correspondence we need some facts on duality. Given A-bimodules M, N, we write $\mathrm{Hom}_{A-}(M,N)$, $\mathrm{Hom}_{-A}(M,N)$, $\mathrm{Hom}_{A,A}(M,N)$ for the set of all left-A, right-A and A-bimodule homomorphisms. Further we put

$$*M = \mathrm{Hom}_{A-}(M,A), \quad M* = \mathrm{Hom}_{-A}(M,A),$$

$$*M* = *M \cap M* = \mathrm{Hom}_{A,A}(M,A).$$

E.g. if M_A is free of finite rank, then $*(M*) \cong M$, as is well known.

Let M be an A-coring, then *M has a ring structure as follows: For f,g ε *M, their product is defined as the composition

$$f.g : M \longrightarrow M \otimes_A M \xrightarrow{1 \otimes f} M \otimes_A A = M \xrightarrow{g} A.$$

Thus $f.g : u \longmapsto \Sigma(u_{i1}u_{i2}^f)^g$ if $\Delta(u) = \Sigma u_{i1} \otimes u_{i2}$. It is easily verified that this is indeed a ring structure. Similarly M* is defined as a ring: the composition of f,g ε M* is defined as

$$f.g : M \longrightarrow M \otimes_A M \xrightarrow{g \otimes 1} A \otimes_A M = M \xrightarrow{f} A.$$

Here f.g maps u to $\Sigma(u_{i1}^g u_{i2})^f$. We note that for f,g ε *M* both definitions reduce to $\Sigma u_{i1}^g u_{i2}^f$, thus both *M and M* contain *M* as subring.

Example. Let $C = B \otimes_A B$ be a standard B-coring, then $C* = \mathrm{Hom}_{-B}(B \otimes_A B, B) = \mathrm{Hom}_{-A}(B, \mathrm{Hom}_{-B}(B,B)) = \mathrm{Hom}_{-A}(B,B)$. Thus

$$C* = \mathrm{End}_{-A}(B),$$

and similarly $*C = \mathrm{End}_{A-}(B)$, while $*C* = \mathrm{End}_{A,A}(B)$. The latter is essentially the centralizer of $\Phi(A)$ in B.

It is clear that if $f:M \longrightarrow N$ is a coring map, then $f*: M* \longrightarrow N*$ and $*f:*M \longrightarrow *N$ are ring homomorphisms. E.g., for any B-coring M, $\varepsilon:M \longrightarrow B$ is a coring map, hence

$$\varepsilon* :B \longrightarrow M*$$

is a ring homomorphism; explicitly, $\varepsilon* :b \longmapsto \lambda_b\varepsilon$, i.e. b corresponds to the map $m \longmapsto \varepsilon(bm)$.

Now let K be any field and End(K) the ring of additive group endomorphisms of K. In End(K) we have the subrings $\rho(K)$, $\lambda(K)$ of right and left multiplications, and as is well known, these are each other's centralizers, thus $\rho(K)=\mathrm{End}_{K-}(K)$ $\lambda(K)=\mathrm{End}_{-K}(K)$. Further, End(K) as subset of K^K can be regarded as a topological space, the topology being induced by the product topology on K^K, taking K with the discrete topology. This is sometimes known as the topology of simple convergence; if $f \in \mathrm{End}(K)$, a typical neighbourhood of f consists of all $\Phi \in \mathrm{End}(K)$ such that for a given finite set $c_1,\ldots,c_n \in K$, $c_if = c_i\Phi$. In particular, this shows every centralizer to be closed.

We shall need one more auxiliary result:

Proposition 3.2.4. *Let* K *be any field. Given a subring* F *of* End(K) *such that* $\rho(K)\subseteq F\subseteq \mathrm{End}(K)$, *define*

$$D = \{x \in K \mid \lambda_x \text{ centralizes } F\}$$

$$= \{x \in K \mid (xy)f = x.yf \text{ for all } y\in K, f\in F\}$$

then (i) $D = \{x \in K \mid xf = x.1f \text{ for all } f \in F\}$, (ii) *the centralizer of* F *in* End(K) *is* $\lambda(D)$ *and hence* D *is a subfield of* K, (iii) $\rho(K)\subseteq F\subseteq \mathrm{End}_{D-}(K)$.

Proof. (i) If $xf = x.1f$, then $(xy)f = x\rho_y f = x.1\rho_y f = x.yf$

because $\rho(K) \subseteq F$. (ii) Since $\rho(K) \subseteq F$, the centralizer of F is contained in $\lambda(K)$; in fact the definition of D states that $\lambda_x f = f\lambda_x$, thus the centralizer is $\lambda(D)$. Now the rest is clear. ∎

Theorem 3.2.5 (Jacobson-Bourbaki). Let K be a skew field and End(K) its endomorphism ring, as topological ring. Then there is an order-reversing bijection between the subfields D of K and the closed $\rho(K)$-subrings F of End(K), defined by the rules

$$(4) \quad D \longmapsto \mathrm{End}_{D-}(K), \qquad F \longmapsto D = \{x \in K \mid xf = x.1f \text{ for } f \in F\}.$$

Further, if D and F correspond, then

$$(5) \quad [K:D]_L = [F:K]_R$$

whenever either side is finite.

Proof. Given F, define D as in (4). Let X be a left D-basis for K and for $x \in X$ define $\delta_x \in \mathrm{End}_{D-}(K)$ by

$$y.\delta_x = \begin{cases} 1 & \text{if } y = x, \\ 0 & \text{if } y \neq x. \end{cases}$$

Then the δ_x are right K-linearly independent, for if $\Sigma\delta_x a_x = 0$ $(a_x \in K)$, we can apply this to $y \in X$ to get $a_y = 0$, so the relation was trivial. Moreover, if $[K:D]_L < \infty$, the δ_x form a basis for $\mathrm{End}_{D-}(K)$ as right K-space, because then $f = \Sigma\delta_x.xf$, as is easily checked. Therefore we have

$$(6) \quad [\mathrm{End}_{D-}(K):K]_R = [K:D]_L$$

if the right-hand side is finite. Now $F \subseteq \mathrm{End}_{D-}(K)$, so

$[F:K]_R$ is finite if $[K:D]_L$ is, by (6). To show that

(7) $$F = \mathrm{End}_{D-}(K),$$

we use Jacobson's density theorem [56] to deduce that F, as $\rho(K)$-subring, is dense in $\mathrm{End}_{D-}(K)$ and being closed is therefore the whole of $\mathrm{End}_{D-}(K)$. In the restricted but still important situation where $[F:K]_R < \infty$, (7) can be proved from Prop. 3.2.2 by verifying that the map

$$\kappa : K \otimes_D K \longrightarrow F^*, \qquad \alpha \otimes \beta \longmapsto (f \longmapsto \alpha\beta^f)$$

is injective. It is of interest that for any $\rho(K)$-subring F the injectivity of κ would follow from the density of F in $\mathrm{End}_{D-}(K)$. Conversely, if

$$*\kappa: \quad *(F^*) \longrightarrow *(K \otimes_D K) = \mathrm{End}_{D-}(K)$$

is surjective, then since F is dense in $*(F^*)$, it follows that F is dense in $\mathrm{End}_{D-}(K)$; thus the injectivity of κ is a predual of the density theorem.

To outline the proof, let $[F:K]_R < \infty$. Then the map

$$\eta: F^* \otimes_K F^* \longrightarrow (F \otimes_K F)^*, \quad \phi\otimes\psi \longmapsto (f \otimes g \longmapsto (f^\psi g)^\phi)$$

is an isomorphism and F^* has a coring structure given by $\varepsilon: F^* \longrightarrow K$, $\phi \longmapsto (1_F)^\phi$, and $\Delta: F^* \xrightarrow{(\text{mult})^*} (F \otimes_K F)^* \xrightarrow{\eta^{-1}} F^* \otimes_K F^*$. Moreover, $*(F^*) \cong F$, as rings. Now $*\kappa$ is an injective ring homomorphism, hence $\kappa = (*\kappa)^*$ is a surjective coring map.

Let $g = g_{K/D} \in K \otimes_D K$ be the standard grouplike and $c = \kappa(g) \in F^*$. Put $D_1 = \{x \in K \mid xc = cx\}$, then $D_1 \supseteq D$; if $b \in K$, $cb = \kappa(1 \otimes b)$ and $bc = \kappa(b \otimes 1)$, hence for any $f \in F$, $(cb)^f = b^f$, $(bc)^f = b.1^f$, so if $b \in D_1$ then $b^f = b.1^f$, i.e. $b \in D$ by Prop. 3.2.4 (i); this shows that $D_1 = D$. Further, $F^* = KcK$, hence κ is an isomorphism by Prop. 3.2.2, therefore so is $*\kappa$, i.e. $F = \mathrm{End}_{D-}(K)$ and $[K:D]_L = [F:K]_R$.

Conversely, given D, put $F = \mathrm{End}_{D-}(K)$ and define $D_1 = \{x \in K \mid xf = x.1f \text{ for all } f \in F\}$, then $D_1 \supseteq D$ and by what we have seen, $F = \mathrm{End}_{D_1-}(K)$, hence $D_1 = D$. Finally if either side of (5) is finite, then by (6) we have

$$[F:K]_R = [K:D]_L$$

as claimed. ■

3.3 Galois theory

Let K be a skew field and G the group of all its automorphisms. For any subfield E of K let $E^+ = \{\sigma \in G \mid x^\sigma = x$ for all $x \in E\}$ and for any subgroup H of G, put $H^+ = \{x \in K \mid x^\sigma = x$ for all $\sigma \in H\}$. Then it is clear that E^+ is a subgroup of G, H^+ is a subfield of K and as in any Galois connexion,

$$E_1 \subseteq E_2 \Rightarrow E_1^+ \supseteq E_2^+, \qquad H_1 \subseteq H_2 \Rightarrow H_1^+ \supseteq H_2^+,$$

$$E \subseteq E^{++}, \qquad H \subseteq H^{++},$$

and hence

$$E^{+++} = E^+, \qquad H^{+++} = H^+.$$

Given a subfield E of K, we call E^+ the *Galois group* of K/E, written Gal(K/E), and given a subgroup H of G, we call H^+ the *fixed field* of H. If $E = H^+$ for some H, we shall say that K/E is *Galois*.

The object of Galois theory is to find which fields in K are of the form $E = H^+$ and which subgroups of Gal(E/K) have the form E^+. We recall that in the case of commutative fields the finite Galois extensions are just the normal separable extensions, while every subgroup of Gal(K/E) has the form F^+ for a suitable field F between K and E. The account which follows is based on Jacobson [56].

The commutative theory rests on two basic results:

Dedekind's lemma. Distinct homomorphisms of a field E *into a field* F *are linearly independent over* F.

Artin's lemma. If G *is a group of automorphisms of a field* E *and* F *is the fixed field, then* $[E:F] = |G|$ *whenever either side is finite.*

Our object is to find generalizations. We begin with Dedekind's lemma; here we have to define what we mean by the linear independence of homomorphisms over a skew field. Given any skew fields K, L, we write $H = \mathrm{Hom}(K,L)$ for the set of all field homomorphisms from K to L. Let HL be the right L-space on the set H as basis and define HL as left K-space by the rule

$$\alpha s = s\alpha^s, \qquad \alpha \in K, \quad s \in H = \mathrm{Hom}(K,L).$$

Thus HL is a (K,L)-bimodule, as is easily checked. Each element of HL defines a mapping $K \longrightarrow L$ as follows

$$(1) \quad \Sigma s_i\lambda_i : \alpha \longmapsto \Sigma\alpha^{s_i}\lambda_i \qquad (\alpha \in K,\ s_i \in H,\ \lambda_i \in L).$$

We observe that for $\alpha,\beta \in K$, $s \in H$, $\alpha^{\beta s} = \alpha^{s\beta^s} = \alpha^s\beta^s = (\alpha\beta)^s$, thus $\alpha^{\beta s} = (\alpha\beta)^s$ and so the left K-module structure of HL acts on K in the expected way. Let N be the kernel of the mapping from HL to $\mathrm{Map}(K,L)$ defined by (1), thus N consists of all sums $\Sigma s_i\lambda_i$ such that $\Sigma\alpha^{s_i}\lambda_i = 0$ for all $\alpha \in K$. Write $M = HL/N$, then M is a (K,L)-bimodule whose elements have the form $\Sigma s_i\lambda_i$ ($s_i \in H, \lambda_i \in L$), with $\Sigma s_i\lambda_i = 0$ if and only if $\Sigma\alpha^{s_i}\lambda_i = 0$ for all $\alpha \in K$.

Each $\mu \in L^*$ defines an inner automorphism of L:

$$I_\mu : \lambda \longmapsto \mu\lambda\mu^{-1},$$

and it is clear that for $s \in H$, $sI_\mu \in H$. Two homomorphisms $s,t: K \longrightarrow L$ are called *equivalent* if they differ by an inner automorphism: $t = sI_\mu$. We note that for each $s \in H$, sL is a

(K,L)-submodule of HL which is simple as (K,L)-module, since it is already simple as L-module. Two homomorphisms s,t define isomorphic (K,L)-bimodules if and only if they are equivalent. For if $sL \cong tL$, say t corresponds to $s\mu$ ($\mu \in L^*$), then for all $\alpha \in K$, $\alpha.s\mu = s\mu.\alpha^t$. Since $\alpha s = s\alpha^s$, we find that $\alpha^s\mu = \mu\alpha^t$, hence s,t are equivalent. Conversely, if s,t are equivalent, then $\alpha^s\mu = \mu\alpha^t$ for some $\mu \in L^*$ and retracing our steps we find that $sL \cong tL$, with t corresponding to $s\mu$ in the isomorphism.

It follows that HL is a sum of simple (K,L)-bimodules, i.e. semisimple, hence so is the quotient M. We recall that a semisimple module is a direct sum of homogeneous components, where each homogeneous component is a direct sum of simple modules of a given type (cf. Jacobson [56] or Cohn [77]). Now we have the following generalization of Dedekind's lemma.

Theorem 3.3.1(i). *Given* $s, s_1, \ldots, s_n \in H = \mathrm{Hom}(K,L)$, *if*

$$s = \Sigma s_i\lambda_i \qquad \text{in } M = HL/N \qquad (\lambda_i \in L),$$

then $s = s_iI_\mu$ *for some* i *and some* $\mu \in L^*$.

(ii) *Given* $s \in H$, *if* $\mu_1, \ldots, \mu_r \in L$ *are such that the elements* sI_{μ_i} *in* M *are linearly dependent over* L, *then the* μ_i *are linearly dependent over* $C(K^s)$, *the centralizer of* K^s *in* L.

Proof. (i) If $s = \Sigma s_i\lambda_i$, then the simple module sL lies in the same homogeneous component as some s_i, so s and s_i generate isomorphic modules, i.e. $s = s_iI_\mu$ ($\mu \in L^*$).

(ii) If the sI_{μ_i} are linearly dependent, take a relation of shortest length:

$$\Sigma_1^p sI_{\mu_i}\lambda_i = 0 \qquad (\lambda_i \in L).$$

Then each $\lambda_i \neq 0$ and by multiplying on the right by a suitable factor we may assume that $\lambda_1 = \mu_1$. Apply this relation to $\alpha\beta \in K$:

$$(2) \qquad 0 = \Sigma(\alpha\beta)^s I_{\mu_i}\lambda_i = \Sigma\mu_i\alpha^s\beta^s\mu_i^{-1}\lambda_i .$$

Next apply the relation to α and multiply by β^s on the right:

$$0 = \Sigma\alpha^s I_{\mu_i}\lambda_i\beta^s = \Sigma\mu_i\alpha^s\mu_i^{-1}\lambda_i\beta^s .$$

Taking the difference, we get

$$\Sigma_1^p\mu_i\alpha^s\mu_i^{-1}\ (\lambda_i\beta^s - \mu_i\beta^s\mu_i^{-1}\lambda_i) = 0.$$

The first coefficient is $\lambda_1\beta^s - \lambda_1\beta^s = 0$, hence by minimality the others are also 0, so $\mu_i^{-1}\lambda_i\beta^s = \beta^s\mu_i^{-1}\lambda_i$, i.e. $\mu_i^{-1}\lambda_i \in C(K^s)$. Now by (2), with $\alpha = \beta = 1$,

$$\Sigma\mu_i . \mu_i^{-1}\lambda_i = 0,$$

and this is the required dependence relation over $C(K^s)$. ■

Corollary 1. *Let* $s_1,\ldots,s_r$ *be pairwise inequivalent isomorphisms between* K *and* L *and let* $\lambda_1,\ldots,\lambda_t \in L$ *be linearly independent over the centre of* L, *then the isomorphisms* $s_i I_{\lambda_j}$ *are linearly independent over* L.

For if they were linearly dependent, then for some $s = s_i$ the sI_{λ_j} would be linearly dependent (because each s_i belongs to a different homogeneous component). So by (ii) the λ_j are linearly dependent over $C(K^s)$, i.e. the centre of L, which contradicts the hypothesis. ■

Corollary 2. *If* $s_1,\ldots,s_r$ *are inequivalent isomorphisms between* K *and* L, $\lambda_1,\ldots,\lambda_t \in L$ *are linearly independent over* C, *the centre of* L *and*

(3) $\quad s = \Sigma s_i I_{\lambda_j} \alpha_{ij} \qquad\qquad (\alpha_{ij} \in L),$

then $s = s_i I_\lambda$, *for some* i, *where* $\lambda = \Sigma\lambda_j\beta_j$ $(\beta_j \in C)$.

Proof. By (i), $s = s_i I_\lambda$ for some i and some $\lambda \in L$. Thus

$$s_i I_\lambda = \Sigma s_i I_{\lambda_j} \alpha_{ij}$$

and by equating homogeneous components we can omit terms s_k with $k \neq i$. Now by Cor.1, $\lambda, \lambda_1, \ldots, \lambda_t$ are linearly dependent over C, but $\lambda_1, \ldots, \lambda_t$ are independent, hence $\lambda = \Sigma\lambda_j\beta_j$, $\beta_j \in C$. ■

Next we have to translate Artin's lemma. Without using Dedekind's lemma the result is $[E:F] = [G:E]$, where G, or rather GE is regarded as right E-space. We shall replace G, a group of F-automorphisms of E, by F-linear transformations of $_F E$. Every such $s \in \mathrm{End}_{F-}(E)$ satisfies

$$(\gamma x)^s = \gamma x^s \qquad\qquad x \in E, \gamma \in F.$$

This generalizes the rules $(xy)^s = x^s y^s$, $\gamma^s = \gamma$ satisfied for $s \in G$. Given any skew field K, we consider the set End(K) of additive group endomorphisms as a topological ring (3.2). This set contains $\rho(K)$, the ring of right multiplications as subring, and we recall the Jacobson-Bourbaki correspondence (p.38):

There is an order-reversing bijection between the subfields D of K and the closed $\rho(K)$-subrings F of End(K) such that

$$[K:D]_L = [F:K]_R$$

whenever either side is finite.

Given a group G of automorphisms of K, we have a right

K-space GK, and we need only show that this is a ring in order to be able to apply the preceding result. Thus we have

Proposition 3.3.2. *Let* K *be any skew field and* G *a group of automorphisms of* K, *then* GK *is a* $\rho(K)$-*subring of* End(K) *and its closure* $\overline{GK}$ *is* $\mathrm{End}_{D-}(K)$, *where* D *is the subset of* K *left fixed by* G.

Proof. In GK we have the rule $\alpha g = g\alpha^g$ $(\alpha \in K,\ g \in G)$. Using this rule, we have

$$(g_1\alpha_1)(g_2\alpha_2) = g_1g_2\alpha_1^{g_2}\alpha_2.$$

Since every element of GK is a sum of terms $g\alpha$, it follows that GK is closed under products and contains 1, hence it is a ring, indeed a $\rho(K)$-ring, because $\rho(K) \subseteq GK$. By the Jacobson-Bourbaki correspondence, $\overline{GK} = \mathrm{End}_{D-}(K)$, where $D = \{x \in K \mid x^f = x.1^f \text{ for all } f \in \overline{GK}\}$. Thus if $\alpha \in K$, then

$$\begin{aligned} \alpha \in D \iff & \alpha^f = \alpha.1^f \quad \text{for all } f \in \overline{GK}, \\ \iff & \alpha^{g\beta} = \alpha.1^{g\beta} \quad \text{for all } g \in G,\ \beta \in K, \\ \iff & \alpha^g\beta = \alpha\beta, \end{aligned}$$

i.e. $\alpha \in D \iff \alpha^g = \alpha$ for all $g \in G$. ■

Combining this with Th.3.3.1, we get

Proposition 3.3.3. *Let* K *be a skew field,* G *a group of automorphisms of* K, *and* D *the fixed field of* G. *Assume that* G *contains every inner automorphism of* K *over* D. *If* E *is a subring of* K *such that* $D \subseteq E \subseteq K$ *and* $|E:D|_L < \infty$, *then every* D-*ring homomorphism of* E *into* K *is induced by an element of* G.

Proof. Let $s: E \to K$ be a D-ring homomorphism. Regarding E and K as left D-spaces, we see that E is a subspace, so s can be extended to a D-space endomorphism of K, i.e. an element of $\mathrm{End}_{D-}(K)$. By Prop.3.3.2, $\mathrm{End}_{D-}(K) = \overline{GK}$ and on any finite-dimensional subspace s can be written as $\Sigma g_i\lambda_i$ $(g_i \in G,\ \lambda_i \in K)$. In particular, since $[E:D]_L < \infty$, we have

$$s = \Sigma g_i\lambda_i \quad \text{on } E.$$

Now E is an entire D-ring of finite left degree, hence a field, and by Th.3.3.1 (i), $s = g_i I_\mu$ for some i and some $\mu \in K$. Applying this to $\alpha \in D$ we have

$$\alpha = \alpha^s = \alpha^{g_i I_\mu} = \mu\alpha\mu^{-1}.$$

Thus I_μ is an inner automorphism of K over D, and so $I_\mu \in G$, hence $g_i I_\mu \in G$ is an automorphism which induces s.■

Corollary. Let K *be a skew field with centre* C *and* D *a* C-*subalgebra finite-dimensional over* C, *then any* C-*algebra homomorphism* D ⟶ K *can be extended to an inner automorphism of* K.

This follows by taking G to be the group of all inner automorphisms of K, then C is the fixed field and every inner automorphism belongs to G, so the proposition may be applied.

We now return to our initial task of finding which automorphism groups and subfields correspond under the Galois connexion. We first deal with a condition which is obviously satisfied by all Galois groups. Let K be a field with centre C and let D be any subfield of K, then the centralizer of D in K is a subfield D' containing C. Any non-zero element of D' defines an inner automorphism of K which leaves D elementwise fixed and so belongs to the group D^+; conversely, an inner automorphism of K belongs to D^+ only if it is induced by an element of D'. Thus we see that the $\alpha \in K$ for which $I_\alpha \in D^+$ form together with 0 a subfield containing C. This is then a necessary condition for a group to be Galois and we define: A group G of automorphisms of K is called an N-*group* (after E.Noether) if the set

$$A = \{\alpha \in K \mid \alpha = 0 \text{ or } I_\alpha \in G\}$$

is a C-subalgebra of K. We shall call this the C-algebra *associated* with G.

Clearly the associated C-algebra is necessarily a field. If G is any N-group with associated algebra A, and G_o is the subgroup of inner automorphisms I_α ($\alpha \in A$), then G_o is normal in G, for if $x \in K$, $\alpha \in A^*$, $s \in S$, then $x^{s^{-1}I_\alpha s} = (\alpha x^{s^{-1}}\alpha^{-1})^s = \alpha^s x(\alpha^s)^{-1} = xI_{\alpha^s}$, so

$$s^{-1}I_\alpha . s = I_{\alpha^s}.$$

We define the *reduced order* of G as

$$|G|_{red} = (G:G_o)\left[A:C\right].$$

With this notation we have

Theorem 3.3.4. *Let* K *be any skew field,* G *an* N-*group of automorphisms of* K *and put* $D = G^+$. *Then*

$$(4) \qquad \left[K:D\right]_L = |G|_{red}$$

whenever either side is finite, and when this is so, $G^{++} = G$.

Proof. Suppose first that $\left[K:D\right]_L = m < \infty$, then by Th.3.2.5 (J.-B.) $\left[End_{D-}(K):K\right]_R = m$. Let $s_1,\ldots,s_r \in G$ be pairwise incongruent (mod G_o) and $\lambda_1,\ldots,\lambda_t$ any elements of the associated C-algebra A that are linearly independent over C. Then the maps $s_i I_{\lambda_j}$ are in $End_{D-}(K)$ and by Th.3.3.1, Cor.1 they are linearly independent over K; hence $rt \leq m$ and it follows that $|G|_{red} < \infty$.

We may now assume that $|G|_{red}$ is finite. Let $s_1,\ldots,s_r$ be a transversal for G_o in G and $\lambda_1,\ldots,\lambda_t$ a C-basis for A, then we know that the $s_i I_{\lambda_j}$ are right linearly independent over K. We shall show that they form a basis for $End_{D-}(K)$.

Let $s \in G$, then $s = s_1 I_\lambda$ say, where $\lambda \in A$ and so $\lambda =$

$\Sigma\lambda_j\beta_j$ ($\beta_j \in C$). For any $c \in K$ we have

$$\begin{aligned} c^s = c^s 1 I_\lambda &= \lambda c^s 1 \lambda^{-1} \\ &= \Sigma\lambda_j\beta_j c^s 1 \lambda^{-1} \\ &= \Sigma\lambda_j c^s 1 \lambda_j^{-1} . \lambda_j \beta_j \lambda^{-1} \\ &= \Sigma c^s 1 I_{\lambda_j} \gamma_j , \end{aligned}$$

where $\gamma_j = \lambda_j\beta_j\lambda^{-1} \in K$. This shows that the $s_i I_{\lambda_j}$ span G, hence also GK. Now GK has finite dimension over K and so $GK = \overline{GK} = \mathrm{End}_{D-}(K)$ (Prop.3.3.2). It follows that $|G|_{red} = [\mathrm{End}_{D-}(K):K]_R = [K:D]_L$.

Next it is clear that $D^+ = G^{++} \supseteq G$. Conversely, if $s \in D^+$, then $s \in \mathrm{End}_{D-}(K)$, hence $s = \Sigma s_i I_{\lambda_j} \alpha_{ij}$. By Th.3.3.1, Cor. 2, since s is an automorphism of K, $s = s_i I_\lambda$ for some i, where $\lambda = \Sigma\lambda_j\beta_j$ ($\beta_j \in C$), but then $I_\lambda \in G_o$ and $s_i \in G$, hence $s \in G$. ■

Here is an example, taken from Amitsur [54], to illustrate the need for introducing the reduced order.

Let F be the $\mathbf{Q}$-algebra generated by u, with defining relation

$$u^2 + u + 1 = 0.$$

F admits a $\mathbf{Q}$-automorphism $\sigma: u \longmapsto u^2$. Put $R = F[v;\sigma]$, then $v^2 - 2$ is central and irreducible, because the equation

$$2 = (x + yu)(x + yu^2) = x^2 - xy + y^2$$

has no rational solution. So we can form the skew field $K = R/(v^2 - 2)$. Now the group generated by I_u has order 3, but reduced order 2, for its fixed field is F and $[K:F] = 2$.

We note some consequences of Th.3.3.4.

Corollary 1. *If* K/D *is Galois (i.e.* $D = G^+$ *for some* G*), then*

$$[K:D]_L = [K:D]_R$$

whenever either side is finite.

This follows from (4) by symmetry. ∎

Corollary 2. *Let* K *be a skew field with centre* C *and* A *any* C-*subalgebra of* K. *Then the centralizer* A' *of* A *is again a* C-*subalgebra of* K *and* $A'' \supseteq A$. *Moreover,*

$$(5) \quad [K:A']_L = [A:C]$$

whenever either side is finite, and when this is so, $A'' = A$.

Proof. Suppose first that $[K:A']_L$ is finite. Clearly A' is a subfield of K; let G be the group of inner automorphisms of K fixing A' and A_1 the associated algebra, then $A_1 \supseteq A$ and by Th.3.3.4, $|G|_{red} = [A_1:C] = [K:A']_L$, hence A is then finite-dimensional. Thus we may assume that $[A:C]$ is finite, hence A is a field. Now let G be the group of inner automorphisms induced by A, then clearly $G^+ = A'$, and (5) follows from (4). Moreover, $(A')^+ = G$, which means that $A'' = A$. ∎

Corollary 3. *If* K/D *is Galois with group G and* $[K:D]_L < \infty$, *then any* $\rho(K)$-*subring* B *of* GK *has the form* HK, *where* $H = G \cap B$ *is an* N-*subgroup of* G.

Proof. Let $H = G \cap B$, then clearly $HK \subseteq B$. To prove equality we note that HK and B are both K-bimodules contained in $GK = \mathrm{End}_{D-}(K)$, by Prop.3.3.2. Now GK is semisimple and hence so is B; moreover every simple submodule M of B is isomorphic to a simple submodule of GK and hence is of the form $M = uK$, where $\alpha u = u\alpha^s$ for all $\alpha \in K$ and some $s \in G$. Replacing u by $u\gamma$ ($\gamma \in K$) if necessary, we may suppose that

$1.u = 1$ and still $\alpha u = u\alpha^s$ (for some $s \in G$).

Hence $\alpha.u = 1.u\alpha^s = \alpha^s$, i.e. u is an automorphism of K, viz. s, and since $u \in End_{D^-}(K)$, s fixes D, i.e. $s \in G$, so $s \in G \cap B =$ H; this shows that B = HK.

That H is a group follows because B is a centralizer (bein finite-dimensional over K, by Jacobson-Bourbaki). To show that H is an N-group, let $I_{\alpha_1}, I_{\alpha_2} \in H$ and $\alpha = \alpha_1\beta_1 + \alpha_2\beta_2 \neq 0$; we must show that $I_\alpha \in H$ and clearly it will be enough to show that $I_\alpha \in B$. Now $I_\alpha = \lambda(\alpha)\rho(\alpha^{-1})$, and $B \supseteq \rho(K)$, hence $I_\alpha \in B \iff \lambda(\alpha) \in B$. By hypothesis, $\lambda(\alpha_1), \lambda(\alpha_2) \in B$ and $\lambda(\alpha) = \lambda(\alpha_1)\beta_1 + \lambda(\alpha_2)\beta_2$ hence $\lambda(\alpha) \in B$. ■

Finally we come to the fundamental theorem. In order to describe extensions, we need an analogue of normal subgroups, bearing in mind that we admit only N-groups. We therefore define: A subgroup H of an N-group G is *N-invariant* in G if the N-subgroup generated by the normalizer of H in G is G itself.

Theorem 3.3.5 *(Fundamental theorem of Galois theory for skew fields). Let* K/F *be Galois with group G and* $[K:F]_L < \infty$.

(i) *There is a bijection between* N-*subgroups of* G *and fields* D, $F \subseteq D \subseteq K$:

$$H \longmapsto H^+ = \{x \in K \mid x^\sigma = x \text{ for all } \sigma \in H\},$$
$$D \longmapsto D^+ = \{\sigma \in G \mid x^\sigma = x \text{ for all } x \in D\}.$$

If $H \longleftrightarrow D$, *then* K/D *is Galois with group* H *and* $[K:D]_L = |H|_{red}$.

(ii) *If* $H \longleftrightarrow D$, *then the group of automorphisms of* D/F *is isomorphic to* N_H/H, *where* N_H *is the normalizer of* H *in* G. *Moreover,* D/F *is Galois if and only if* H *is* N-*invariant in* G.

Proof. (i) By Th.3.3.4, $|G|_{red} < \infty$ and so any N-subgroup of G has finite reduced order: $H/H_o = H/H \cap G_o \cong HG_o/G_o \subseteq G/G_o$, and the C-algebra associated with H is contained in that of G. Hence, given H, we have by Th.3.3.4, $H^{++} = H$. Conversely, given D, put $H = G \cap \mathrm{End}_{D^-}(K)$, then $\mathrm{End}_{D^-}(K) = HK$, by Th.3.3.4, Cor.3. Now H consists of all D-linear transformations in G, i.e. the elements of G fixing D, thus $H = D^+$. Since $\mathrm{End}_{D^-}(K) = HK$, we have $D = H^+$ by Prop.3.3.2. Thus $D^{++} = D$, and now $[K:D]_L = |H|_{red}$ by Th.3.3.4.

(ii) Let $D \longleftrightarrow H$, then for any $s \in G$, $D^s \longleftrightarrow s^{-1}Hs$, therefore the elements of N_H and only these are automorphisms of D/F. Every automorphism of D/F is induced by an automorphism of K/F, hence its Galois group, G_1 say, is a homomorphic image of N_H. The kernel consists of those automorphisms of K which fix D, i.e. H, so that

$$G_1 \cong N_H/H.$$

So far D was any field between K and F. Now D/F is Galois if and only if F is the fixed field of G_1, i.e. F is the set of elements fixed by N_H. By (i) this holds if and only if the N-subgroup of G generated by N_H corresponds to F, which happens precisely when this group is G. But this just means that H is N-invariant in G. ∎

We note the special case where the Galois group is inner (i.e. consists entirely of inner automorphisms). Let L/K be an extension with inner Galois group G and let K' be the centralizer of K in L, then $C = K \cap K'$ is the centre of K and $G \cong K'^*/C^*$. Such extensions are described by Th.3.3.4, Cor.2.

Returning to the case of a general Galois extension L/K with group G, denote by G_o the subgroup of inner automorphisms. As we have seen, G_o is normal in G; moreover,

G/G_o consists entirely of outer automorphisms (apart from 1). For if L_o is the fixed field of G_o and $\bar{\sigma} \in G/G_o$, suppose that $\bar{\sigma} = I_c$ and take $\sigma \in G$ such that $\sigma \mapsto \bar{\sigma}$, then $\sigma^{-1} I_c$ fixes L_o and so lies in G_o, i.e. $\sigma \in G_o$ and hence $\bar{\sigma} = 1$. Let us call a Galois group *outer* if the only inner automorphism it contains is 1. E.g. in the commutative theory all Galois groups are outer. Now we have the

Corollary. Given any Galois extension L/K, *there is a field* L_o, $K \subseteq L_o \subseteq L$, *such that* L_o/K *is Galois with outer Galois group and* L/L_o *has inner Galois group.* ∎

Briefly: *every Galois extension is an outer extension followed by an inner extension.*

Besides automorphisms one can also consider antiautomorphisms of a skew field (reversing the multiplication). If a skew field K has an antiautomorphism, then the set of all automorphisms and antiautomorphisms forms a group in which the automorphisms form a subgroup of index ≤ 2. In th case the set E of fixed elements need not be a subfield, but it is closed under addition and as far as multiplication can be carried out in E it is commutative.

The 'theorem of the primitive element' has an analogue in the skew case:

Proposition 3.3.6. *If* L/K *is an extension of finite degree, with finitely many fields between* L *and* K, *then it is simple, i.e. generated over* K *by a single element.*

Proof. If K is finite, then so is L. Then K,L are commutative and the result is well known. In general take a simple extension K(a) of maximal degree. If $K(a) \neq L$, take $b \in L$, $b \notin K(a)$ and consider the fields $L_\lambda = K(a + \lambda b)$ $(\lambda \in K)$. Two of these must coincide because K is infinite; if L_γ contains $a + \gamma b$, $a + \gamma' b$, then it also contains $(\gamma - \gamma')b$, hence b and a, thus L_γ is a simple extension strictly larger than K(a), a contradiction. ∎

The conclusion holds for any Galois extension with finite

Galois group G. However, a Galois group of finite reduced order may well have infinitely many N-subgroups.

In a finite Galois extension (or any finite extension, for that matter) every element is right algebraic over the ground field K. In these circumstances we have

Proposition 3.3.7. *Let* L/K *be any Galois extension (not necessarily finite), with group* G. *If* $a \in L$ *is right algebraic over* K *with minimal polynomial* $p(t)$, *then*

$$p(t) = (t - a_1)(t - a_2)\dots(t - a_n), \tag{6}$$

where $a_1,\dots,a_n$ *are conjugates of* G-*transforms of* a, *and every* G-*transform of* a *is conjugate to one of* $a_1,\dots,a_n$.

Proof. Clearly $p(t) = (t - a)f(t)$ in $L[t]$, and if $\sigma \in G$, then $p(t) = (t - a^\sigma)f^\sigma(t)$, hence all the $t - a^\sigma$ have a least common right multiple $q(t)$ which is a left factor of $p(t)$. Now $q(t)$ is invariant under G and so has coefficients in K, therefore $q = p$. Let us write

$$p(t) = (t - a_1)\dots(t - a_r)p_2 = p_1p_2,$$

where $a_1 = a$, each a_i is a conjugate of a G-transform of a and r is chosen as large as possible. Then $r \geq 1$, because $a_1 = a$; we claim that $p_2 = 1$. If not, there exists a G-transform a' of a such that $t - a'$ is not a left factor of p_1, but p_1 and $t - a'$ have a common right multiple p, hence their least common right multiple can be written in the form $(t - a')p' = p_1(t - a_{r+1})$. Here $t - a_{r+1}$ is similar to $t - a'$, i.e. $(t - a')c' = c(t - a_{r+1})$ for some $c, c' \in L^*$. Equating coefficients we see that $c' = c$, $a'c = ca_{r+1}$, so a_{r+1} is conjugate to a'. Now $(t - a')p'$ is a least common right multiple of $t - a'$ and p_1, and $p = (t - a')p_o = p_1p_2$, therefore we have $p_o = p'q$, $p_2 = (t - a_{r+1})q$, and this contradicts the choice of r. This proves (6); now any $t - a^\sigma$ is a left factor of p and by the same argument is conjugate to one of the a_i. ■

In particular we obtain the well known result that the zeros of p(t) of degree n lie in at most n conjugacy classes (cf. Gordon-Motzkin [65], also Cohn [71''] p.301). Taking L to be any skew field and G the group of inner automorphisms, we find that K is the centre of L and Prop.3.3.7 tells us that the zeros of any polynomial irreducible over the centre are conjugate (L.E. Dickson's theorem).

We conclude this section with a crossed product representation of arbitrary skew fields which is sometimes useful. But let us first recall the usual crossed product construction.

Given a finite Galois extension of commutative fields, F/k, put $\Gamma = \mathrm{Gal}(F/k)$, then each element c of $H^2(\Gamma, F^*)$ defines a crossed product as follows. Let $\{u_\sigma\}$ be a family of elements indexed by Γ and let A be the right F-space on the u_σ as basis; A may be defined as F-bimodule by the equations

$$(7) \qquad au_\sigma = u_\sigma a^\sigma \qquad\qquad (a \in F\ ,\ \sigma \in \Gamma).$$

Now a ring structure may be defined on A by the equations

$$(8) \qquad u_\sigma u_\tau = u_{\sigma\tau} c_{\sigma,\tau}$$

where $c_{\sigma,\tau}$ is a cocycle (factor set) representing $c \in H^2(\Gamma, F^*)$. Then A is a central simple k-algebra, and this is what one usually understands by a crossed product. Conversely, if a central simple k-algebra A has a maximal commutative subfield F which is a splitting field of A and is Galois over k, then A is a crossed product (cf.e.g. Cohn [77]). It turns out that it is not necessary to assume F maximal; if we drop this assumption (but restrict to the case of central division algebras for simplicity) we get

Theorem 3.3.8. *Let* D *be a skew field with centre* k *and let* F *be a commutative subfield containing* k, *such that* F/k *is a finite Galois extension, then* D *is a crossed product over* F', *the centralizer of* F *in* D, *with group* Gal(F/k).

Proof. We observe that F' is a field and by Th.3.3.4, Cor.2, F" = F, so F is the centre of F'. Let U be the normalizer of F* in D*. Any element $u \in U$ induces an automorphism of F/k, hence we have a mapping $\phi: U \longrightarrow \Gamma = \mathrm{Gal}(F/k)$, which is clearly a homomorphism. By the Skolem-Noether theorem it is surjective and the kernel is the centralizer of F* in D*, i.e. F'*. Thus we have a short exact sequence

$$1 \longrightarrow F'^* \longrightarrow U \longrightarrow \Gamma \longrightarrow 1.$$

Let $\{u_\sigma\}$ be a transversal of Γ in U and let D_1 be the right F'-space spanned by the u_σ. It is easily checked that $c_{\sigma,\tau} = u_{\sigma\tau}^{-1} u_\sigma u_\tau \in F'$, hence D_1 is a crossed product over F' with group Γ. The linear independence of the u_σ over F' follows by the familiar argument: if

$$\Sigma u_\sigma a_\sigma = 0 \qquad (a_\sigma \in F')$$

is a shortest non-trivial relation with $a_1 \neq 0$, say, then for any $b \in F$, $\Sigma u_\sigma (b^\sigma - b) a_\sigma = 0$ is a shorter relation, and for suitable b is non-trivial, a contradiction.

Now $[D:F']_R = [F:k]$ by Th.3.3.4, Cor.2, hence $[D:F'] = [D_1:F']$ and so $D_1 = D$, thus D has been expressed as a crossed product over F' with group $\Gamma = \mathrm{Gal}(F/k)$. ■

3.4 Pseudo-linear extensions

It seems hopeless at present to describe all extensions

of finite degree. We shall therefore concentrate on a particular class in this section.

We begin by looking at *quadratic* extensions, i.e. extensions L/K, where $[L:K]_R = 2$. Then for any $a \in L$, $a \notin K$, the pair 1,a is a right K-basis of L. Since every element of L has a unique expression of the form $a\alpha + \beta$ ($\alpha,\beta \in K$), we have in particular,

$$(1) \quad \alpha a = a\alpha^S + \alpha^D \qquad \text{for all } \alpha \in K,$$

and

$$(2) \quad a^2 + a\lambda + \mu = 0 \quad \text{for certain } \lambda,\mu \text{ in } K.$$

Here α^S, α^D are uniquely determined by α and a calculation as in 1.3 shows that S is an endomorphism of K and D an S-derivation. Moreover, the structure of L is completely determined by K and (1),(2).

Conversely, let K be any (skew) field with endomorphism S and S-derivation D. In $R[x;S,D]$ consider a quadratic polynomial $f = x^2 + x\lambda + \mu$. It is easy to write down conditions on λ,μ for f to be right invariant, i.e. for $Rf \subseteq fR$, so that fR is a two-sided ideal, and when these conditions hold, L = R/fR is a K-ring of right degree two. It will be a field precisely if the equation (2) has no solution $a \in K$. We shall not carry out the details here (cf. Cohn [61']), but generalize by keeping (1) and modifying (2).

Thus we define a *pseudo-linear extension* of right degree n, with generator a, as an extension field L of K with right K-basis $1,a,\ldots,a^{n-1}$ such that (1) holds and (in place of (2)):

$$(3) \quad a^n + a^{n-1}\lambda_1 + \ldots + a\lambda_{n-1} + \lambda_n = 0 \qquad (\lambda_i \in K).$$

The remarks made earlier show that every quadratic extension is pseudo-linear. This does not remain true for extensions of higher degree.

We have the following formula for the left degree of a pseudo-linear extension of right degree n:

$$(4)\qquad [L:K]_L = 1 + [K:K^S]_L + [K:K^S]_L^2 + \dots + [K:K^S]_L^{n-1} .$$

In particular, this formula shows that for pseudo-linear extensions

$$[L:K]_L \geq [L:K]_R,$$

with equality if and only if S is an automorphism of K. To prove (4) let us write $L_o = K$, $L_i = aL_{i-1} + K$ $(i \geq 1)$, then $L_i = K + aK + \dots + a^iK$, by an easy induction. Hence we have a chain

$$K = L_o \subset L_1 \subset \dots \subset L_{n-1} = L,$$

and (4) will follow if we can show that each L_i is a left K-space and

$$(5)\qquad \dim_K(L_i/L_{i-1}) = [K:K^S]_L^i .$$

That L_i is a left K-space is clear by pseudo-linearity (using induction). Let $\{u_\lambda\}$ be a left K^S-basis for K; we claim that the elements

$$(6)\qquad a^i u_{\lambda_{i-1}}^{S^{i-1}} \dots u_{\lambda_1}^S u_{\lambda_o}$$

(where $\lambda_o, \lambda_1, \dots, \lambda_{n-1}$ range independently over the index set used for $\{u_\lambda\}$) form a basis of L_i (mod L_{i-1}). This will prove (5) and hence (4). For any $\alpha \in K$ we have

$$\alpha = \Sigma\alpha^S_{\lambda_o} u_{\lambda_o} = \Sigma\alpha^{S^2}_{\lambda_o\lambda_1} u^S_{\lambda_1} u_{\lambda_o} = \ldots = \Sigma\alpha^{S^i}_{\lambda_o\ldots\lambda_{i-1}} u^{S^{i-1}}_{\lambda_{i-1}} \ldots u_{\lambda_o} .$$

Therefore

$$a^i\alpha = \Sigma a^i\alpha^{S^i}_{\lambda_o\ldots\lambda_{i-1}} u^{S^{i-1}}_{\lambda_{i-1}} \ldots u_{\lambda_o}$$

$$\equiv \Sigma\alpha_{\lambda_o\ldots\lambda_{i-1}} a^i u^{S^{i-1}}_{\lambda_{i-1}} \ldots u_{\lambda_o} \quad (\text{mod } L_{i-1}),$$

and this shows that the elements (6) span L_i (mod L_{i-1}). To prove their independence, let

$$\Sigma\alpha_{\lambda_o\ldots\lambda_{i-1}} a^i u^{S^{i-1}}_{\lambda_{i-1}} \ldots u_{\lambda_o} \equiv 0 \quad (\text{mod } L_{i-1}),$$

then on retracing our steps we find that all coefficients

$$\alpha_{\lambda_o\ldots\lambda_{i-1}} = 0 ,$$

hence the elements (6) are linearly independent and so form a basis of L_i (mod L_{i-1}). This establishes (4).

As in the case of quadratic extensions it is clear that every pseudo-linear extension is of the form R/fR, where $R = K[x;S,D]$ and f is a right invariant polynomial, which is irreducible over K. Let us determine all right invariant polynomials in the special case $D = 0$ (for the case where D is arbitrary and S is an automorphism, see Cohn [71"] p.297 and [c]).

Proposition 3.4.1. Let K *be a skew field and* S *an endomorphism of* K *and consider the skew polynomial ring* $K[x;S]$.

(i) *If no power of* S *is inner on* K, *then the right in-*

variant elements are all of the form $x^n\alpha$ $(\alpha \in K)$.

(ii) *If* S^r *is inner on* K, *but no lower power of* S *fixes the centre of* K, *then there exists* $e \in K^*$ *such that* $x^r e = y$ *centralizes* $K[x;S]$ *and the right invariant elements are all of the form* $x^n f\alpha$, *where* f *is a polynomial in* y *with coefficients in the centre of* K *and fixed under* S, *and* $\alpha \in K$.

Proof. Let $f \in K[x;S]$ be right invariant, then so is $f\alpha$, for any $\alpha \in K$, so we may as well take f to be monic, say

$$f = x^n + x^{n-1}\lambda_1 + \dots + \lambda_n \qquad (\lambda_i \in K).$$

By hypothesis, $xf = fg$ for some $g \in K[x;S]$. Comparing degrees we see that g is linear, comparing highest terms we see that g is monic, and comparing lowest terms we see that g vanishes for $x = 0$, hence $g = x$ and so $xf = fx$. If we write this out, we get the equations

$$(7) \qquad \lambda_i^S = \lambda_i \qquad (i = 1,\dots,n).$$

Next we have $\alpha f = fg$, and here g is of degree 0, in fact $g = \alpha^{S^n}$, by comparing highest terms. Equating coefficients, we find $\alpha^{S^{n-i}}\lambda_i = \lambda_i\alpha^{S^n}$; if we use (7) and cancel S^{n-i}, we obtain

$$(8) \qquad \alpha\lambda_i = \lambda_i\alpha^{S^i} \qquad (i = 1,\dots,n).$$

Suppose now that no power of S is inner, then $\lambda_i = 0$ by (8), so $f = x^n$ and every right invariant element has the form $x^n\alpha$ $(\alpha \in K)$. If S^r is inner, say $S^r = I_e$ $(e \in K^*)$, then S^r leaves the centre of K fixed, but no lower power does, so the only powers of S that are inner are multiples of r. By (8) we see that $\lambda_i = 0$ unless $i \equiv 0 \pmod r$; we can therefore write f as x^m times a polynomial in x^r, or equivalently a polynomial in $y = x^r e$, say $f = x^m g$. By definition of e we have

$$(9) \qquad \alpha^{S^r} = e\alpha e^{-1} \qquad \text{for all } \alpha \in K.$$

If we apply S to (9) and compare this with the result of replacing α by α^S in (9) we get $e^S \alpha^S (e^{-1})^S = e\alpha^S e^{-1}$, hence $e^{-1}e^S = \lambda$ commutes with all α^S, i.e. λ lies in the centre C of K. Now S induces an automorphism of C, of order r by hypothesis. By (9) $e^{S^r} = e$, hence

$$\lambda\lambda^S \ldots \lambda^{S^{r-1}} = e^{-1}e^S.e^{-S}e^{S^2}\ldots e^{-S^{r-1}}e^{S^r} = 1.$$

By Hilbert's theorem 90, applied to C (cf.3.5), there exists $\mu \in C$ such that $\lambda = \mu^S\mu^{-1}$, and if we put $e_1 = e\mu^{-1}$, then $x^r e_1$ centralizes K as before, while $x^r e_1 x = x^r e\mu^{-1}x = x^{r+1}e^S\mu^{-S} = x^{r+1}e\mu^S\mu^{-1}\mu^{-S} = x^{r+1}e_1$. Thus we can choose e so that $y = x^r e$ lies in the centre of $K[x;S]$. We saw that any right invariant polynomial has the form $x^m g$, where g is a polynomial in y, which may again be taken monic. Thus let

$$f = x^m(y^t + y^{t-1}\mu_1 + \ldots + \mu_t),$$

then $xf = fx$, hence $\mu_i^S = \mu_i$, and $\alpha f = f\beta$, where $\beta = \alpha^{S^m}$ by a comparison of leading terms, and $\beta\mu_i = \mu_i\beta$, i.e. $\mu_i \in C$. Thus f is of the required form and it is clear that every f of this form is right invariant.∎

The remaining case, when a power of S is inner but a lower power fixes the centre, is more complicated and we shall have no more to say about it. The above result may be stated by saying that under the given conditions the right invariant elements of $R = K[x;S]$ are products of a power of x, a centra element of R and a unit. Since every central element of R is invariant, we obtain the

Corollary. Let K *be a skew field and* S *an endomorphism of* K. *If no power of* S *is inner on* K, *then the centre of* $K[x;S]$ *is the subset of the centre of* K *fixed by* S. *If* S^r *is inner but no lower power fixes the centre of* K, *then the centre of* $K[x;S]$ *has the form* $C_o[y]$, *where* $y = x^r e$ *is central and* C_o *is the subset of the centre of* K *fixed by* S. ■

Let us call a pseudo-linear extension *central* if the associated endomorphism is 1 (and the derivation is 0), and the generator satisfies a monic equation with coefficients in the centre C of K. Thus a central extension of K is of the form $K \otimes_C F$, where F is a commutative field extension of C. Secondly we define a *binomial* or *pure* extension as a pseudo-linear extension in which the generator satisfies a binomial equation

$$x^n - u = 0,$$

and whose associated endomorphism S is such that S^n is the inner automorphism induced by u.

Theorem 3.4.2. *Let* K *be a skew field with endomorphism* S. *Assume that the least power* S^r *which is inner on* K *is also the least power leaving the centre* C *elementwise fixed. Then every pseudo-linear extension of* K *with endomorphism* S *(and zero derivation) is obtained by taking a central extension, followed by a binomial extension.*

Proof. As already remarked, it is clear that every pseudo-linear extension has the form R/fR, where $R = K[x;S]$ and f is a right invariant polynomial in R which is irreducible. If no power of S is inner, there are no such extensions, by Prop.3.4.1. If S^r is inner but no lower power fixes the centre, let $y = x^r e$ be central in R, as in the proof of Prop.3.4.1, then any irreducible right invariant monic polynomial (other than x) is of the form g(y), where g has coefficients in C_o, the subset of C fixed by S. Let F = $C[y]/(g)$ be the commutative field extension defined by g, then $L = K \otimes_C F$ is a central extension of K and the given extension is a binomial extension of L (to which S has been extended by the identity on F), with the defining equation

$$x^r - ye^{-1} = 0. \blacksquare$$

If we are looking among pseudo-linear extensions for examples with different left and right degrees we can concentrate on binomial extensions, by this result. Although as a matter of fact, we shall need to have non-zero derivations too, this suggests that we begin by looking at binomial extensions; our aim will be a result which provides us with a supply of them. First a lemma:

Lemma 3.4.3. *Let* p *be a prime number and* ω *a primitive root of* 1. *If* u,v *are indeterminates over* $\mathbf{Z}[\omega]$ *such that*

$$(10) \quad vu = \omega uv,$$

then

$$(11) \quad (u + v)^p = u^p + v^p.$$

This formula also holds in characteristic p, *with* $\omega = 1$.

Proof. The last part is clear: it is well known that (11) holds in characteristic p if u and v commute. In characteristic $\neq p$ we expand $(u + v)^p$ and consider the terms of degree i in u: they are u^iv^{p-i} together with the terms obtained from it by shifting a u past a v, one at a time. By (10) we can shift it back, inserting a factor ω; thus we get

$$u^i v^{p-i}(1 + \omega + \omega^2 + \ldots + \omega^\nu), \text{ where } \nu = \binom{p}{i} - 1.$$

Since $p \mid \binom{p}{i}$ for $0 < i < p$ and $1 + \omega + \ldots + \omega^{p-1} = 0$, it follows that the sum written is 0, for $0 < i < p$, so all terms in the expression of $(u + v)^p$ vanish except the first and last, i.e. (11). ■

We can now describe a particular class of binomial extensions.

Theorem 3.4.4. *Let* p *be a prime number and* E *a skew field with an endomorphism* S *and a primitive* pth *root of* 1, ω *say, in the centre of* E *and fixed by* S *(of course if char* $E = p$, *then* ω *must be* 1). *Let* D *be an* S-*derivation of* E *such that*

$$\text{(12)} \quad DS = \omega SD.$$

Put $L = E(t;S,D)$, *then* S *may be extended to an endomorphism of* L, *again written* S, *and* D *to an* S-*derivation, by putting*

$$\text{(13)} \quad t^S = \omega t, \qquad t^D = (1 - \omega)t^2,$$

and with these definitions

$$\text{(14)} \quad ct = tc^S + c^D \qquad \textit{for all } c \in L.$$

Moreover, $\sigma = S^p$ *is an endomorphism of* L, $\delta = D^p$ *is a* σ-*derivation, and if* K *is the subfield of* L *generated by* $\tau = t^p$ *over* E, *then* $K = E(\tau;\sigma,\delta)$ *and* L/K *is a binomial extension of degree* p.

Proof. We first observe that ω lies in the centre of L. For characteristic p this is clear, otherwise we have $\omega^p = 1$, hence $0 = (\omega^p)^D = p\omega^{p-1}\omega^D$ and so $\omega^D = 0$. By hypothesis $\omega^S = \omega$, so $\omega t = t\omega$ and the conclusion follows.

To prove (14) we note that it holds for $c \in E$ by definition. Next let $c = t^n\alpha$, where $\alpha \in E$. Then by (13),

$$(t^n\alpha)^S = \omega^n t^n \alpha^S,$$

and (remembering that $t^D = (1 - \omega)t^2$), we have

$$\begin{aligned}(t^n\alpha)^D &= \Sigma\, t^{\nu-1}.t^D.(t^{n-\nu}\alpha)^S + t^n\alpha^D \\ &= (1 - \omega)t^{n+1}(1 + \omega + \ldots + \omega^{n-1})\alpha^S + t^n\alpha^D \\ &= (1 - \omega^n)t^{n+1}\alpha^S + t^n\alpha^D.\end{aligned}$$

Therefore $t(t^n\alpha)^S + (t^n\alpha)^D = t^{n+1}\alpha^S\omega^n + (1 - \omega^n)t^{n+1}\alpha^S + t^n\alpha^D$ $t^{n+1}\alpha^S + t^n\alpha^D$. On the other hand, $(t^n\alpha)t = t^{n+1}\alpha^S + t^n\alpha^D$, so (14) holds for $c = t^n\alpha$; by addition it holds for all polynomials in t and hence for all fractions, i.e. for all elements of L. Now on $E[t;S,D]$ D is the inner S-derivation induced by t and this still holds for $E(t;S,D)$, once we check that this is consistent with (13): $t^D = t.t - t.t^S = (1 - \omega)t$

Next consider $\sigma = S^p$: this is clearly an endomorphism of E. To show that $\delta = D^p$ is a σ-derivation we rewrite (14) in operator form

$$R_t = L_t S + D,$$

where R_t, L_t indicate right and left multiplication by t and S, D mean: S, D applied to the coefficient (not to t). Then $L_t S = SL_t$ and $D.L_t S = \omega L_t S.D$, hence $R_t^p = (L_t S + D)^p$ $= L_t^p\, S^p + D^p$, by the lemma, i.e.

$$ct^p = t^p c^\sigma + c^\delta \qquad (c \in E),$$

which is the required equation. It now follows that the subfield generated by $\tau = t^p$ is $K = E(\tau;\sigma,\delta)$ and clearly

L/K is a pseudo-linear extension with right K-basis $1,t,\ldots,t^{p-1}$ and $t^p = \tau$, i.e. L/K is binomial. ■

Later in 5.6, we shall use this result to construct extensions of finite right and infinite left degree. The problem will be to choose E,S,D so that $[K:K^S]_L = \infty$.

3.5 Cyclic outer Galois extensions

To illustrate some of the results of this chapter, we shall now determine outer Galois extensions with cyclic group, briefly *outer cyclic extensions*, following Amitsur [48,54].

Let K be a skew field with endomorphism S and S-derivation D. The constants $C = \{a \in K \mid a^D = 0\}$ form a subfield of K. We shall also write a' for a^D and $a^{(n)}$ for a^{Dn}. Consider the "differential equation"

$$(1) \quad p(z) \equiv zp = z^{(n)}a_o + z^{(n-1)}a_1 + \ldots + za_n = 0$$

$$(a_i \in K, a_o \neq 0).$$

Theorem 3.5.1 *(Amitsur* [48]*) The solutions of* (1) *in* K *form a left* C*-space of dimension at most* n.

Proof. Clearly $p(\alpha z) = \alpha p(z)$ for $\alpha \in C$, so the linearity is clear. We shall use induction on n. Suppose first that $a_n = 0$, then $p = \Sigma D^i a_{n-i} = Dq$, where q has degree $n-1$ in D, and by induction $U = \ker q$ has dimension $\leq n-1$. Let $u_1,\ldots,u_r$ be left C-independent solutions of (1), where $u_1 = 1$ without loss of generality, because $a_n = 0$. Then $u_2',\ldots,u_r'$ satisfy $zq = 0$ and are linearly left C-independent, for if $\Sigma_2^r \alpha_i u_i' = 0$, then $\alpha = \Sigma\alpha_i u_i \in C$. By the independence of $u_1 = 1,u_2,\ldots,u_r$ we conclude that $\alpha = \alpha_2 = \ldots = \alpha_r = 0$. Hence $r-1 \leq n-1$ and so $r \leq n$ as claimed.

In the general case ($a_n \neq 0$) let u be any solution of

(1), then $zup = 0$ is an equation of order n in which the coefficient of z is $up = 0$, so its solution space U_o has dimension $\leq n$. Now $\ker p = \{zu \mid z \in U_o\} = U_o u$ and this has the same dimension as U_o. ∎

Let L/K be a cyclic extension of degree n, then

$$[L:K]_L = [L:K]_R \leq |\mathrm{Gal}(L/K)|$$

by Th.3.3.4, with equality provided that the Galois group is outer. We shall here confine ourselves to outer cyclic extensions. Let σ be a generator of the Galois group and write $D = \sigma - 1$, then D is a σ-derivation, for

$$\begin{aligned}(ab)^D = (ab)^\sigma - ab = a^\sigma b^\sigma - ab &= (a^\sigma - a)b^\sigma + a(b^\sigma - b) \\ &= a^D b^\sigma + ab^D.\end{aligned}$$

We also note that the field of D-constants is just the fixed field under σ. By hypothesis, $\sigma^n = 1$, hence

$$h(D) = \Sigma_1^n \binom{n}{\nu} D^\nu = (D + 1)^n - 1 = 0. \qquad (2)$$

Theorem 3.5.2. *Let* L/K *be an outer cyclic extension of degree* n *and assume that* K *contains a primitive* nth *root of* 1, ω *say. Denote by* σ *a generator of the Galois group. Then* L *has a right* K*-basis* $a_1,\ldots,a_n$ *such that*

$$a_\nu^\sigma = \omega^\nu a_\nu \qquad (\nu = 1,\ldots,n). \qquad (3)$$

Proof. L is annihilated by $h(D)$, defined as in (2). Now $h(D) = \sigma^n - 1 = h_1(D)(\sigma - \omega)$, where $h_1(D)$ is of degree $n-1$ in D. By Th. 3.5.1, $\ker h_1(D)$ has dimension $\leq n-1$, hence there exists $a \in L$ such that $a_1 = ah_1(D) \neq 0$. Since $ah(D) = 0$, we have $a_1^\sigma = \omega a_1$. Similarly we can find $a_\nu \neq 0$ with $a_\nu^\sigma = \omega^\nu a_\nu$. To prove that $a_1,\ldots,a_n$ form a basis, assume that $\Sigma a_\nu \alpha_\nu = 0$

($\alpha_\nu \in K$), then $\Sigma\omega^{\nu i} a_\nu \alpha_\nu = \Sigma a_\nu^{\sigma i} \alpha_\nu = 0$, and since $1,\omega,\ldots,\omega^{n-1}$ are all distinct, it follows that the α_ν all vanish (by a Vandermonde type argument). Thus $a_1,\ldots,a_n$ are linearly independent and so form a basis. ∎

The hypothesis on ω can be satisfied whenever n is prime to the characteristic of K. In the next result we shall for simplicity assume that ω is central, but this is not essential.

Theorem 3.5.3. *Let* K *be a skew field with a central primitive* nth *root of* 1, ω, *then there is an outer cyclic extension* L/K *of degree* n *and containing* ω *in its centre if and only if there exists an automorphism* σ *of* K *and* $a \in K$ *such that* (i) $\sigma^n = I_a$, $a\sigma = a$, $\omega\sigma = \omega$, *and no lower power of* σ *is inner,* (ii) $t^n a - 1$ *is irreducible in* $K[t;\sigma]$. *When this is so,* $t^n a - 1$ *is right invariant irreducible in* $R = K[t;\sigma]$ *and* $L = R/(t^n a - 1)R$, *with generating automorphism*

$$\sigma : \Sigma t^\nu c_\nu \longmapsto \Sigma(\omega t)^\nu c_\nu .$$

Proof. (i), (ii) just amount to saying that $t^n a - 1$ is central and irreducible (observe that every right invariant element of R is associated to a central element). So if (i), (ii) hold, we have an outer cyclic extension. Conversely, given an outer cyclic extension, let $x \in L$ be such that $x^\sigma = \omega x \neq 0$ (cf.Th.3.5.2). Then any $c \in K$ satisfies

$$(x^{-1}cx)^\sigma = x^{-1}\omega^{-1}c\omega x = x^{-1}cx,$$

hence $x^{-1}cx \in K$, so $\sigma : c \longmapsto x^{-1}cx$ is an automorphism of K, $\omega\sigma = \omega$ and we have a homomorphism

$$K[t;\sigma] \longrightarrow L \qquad \text{given by } t \longmapsto x,$$

and the generator of the kernel has the form $t^n a - 1$, where a satisfies (i), (ii). ■

If we drop the condition that $\omega\sigma = \omega$, then $\omega\sigma = \omega^{\nu}$ for some ν and it is not hard to write down conditions on ν for an extension to exist. One can also give conditions for an outer cyclic extension of degree n if K merely contains a primitive dth root of 1, where d is a proper factor of n.

As a consequence we have a form of Hilbert's theorem 90:

Corollary. If L/K *is an outer cyclic extension of degree* n *with generating automorphism* σ, *and* $c \in L$, *then the equation*

$$(4) \qquad x^{\sigma} = xc$$

has a non-zero solution in L *if and only if*

$$(5) \qquad cc^{\sigma}\ldots c^{\sigma^{n-1}} = 1.$$

Proof. Clearly $c = a^{-1}a^{\sigma}$ satisfies (5). Conversely, if (5) holds, we have $(\sigma\lambda_c)^n = 1$, where λ_c denotes left multiplication by c, for the left-hand side maps x successively to $x^{\sigma}, cx^{\sigma}, c^{\sigma}x^{\sigma^2}, \ldots, cc^{\sigma}\ldots c^{\sigma^{n-1}}x^{\sigma^n} = x$. Thus we have

$$x\left[(\sigma\lambda_c)^n - 1\right] = 0.$$

This has the form $xp(\sigma)\left[\sigma\lambda_c - 1\right] = 0$ for some polynomial $p(\sigma)$; now $xp(\sigma) = 0$ can be considered as a differential equation (with respect to $D = \sigma - 1$) of order n-1, so its solution space has dimension $\leq n-1$, hence there exists $a \in L$ such that $ap(\sigma) = b \neq 0$, and $b(\sigma\lambda_c - 1) = 0$, i.e. $cb^{\sigma} = b$, so (4) holds for $x = b^{-1}$. ■

In a similar way one can show that $x^{\sigma} = cx$ has a non-zero solution if and only if

$$c^{\sigma^{n-1}}\ldots c^{\sigma}c = 1.$$

We also note the following criterion for reducibility:

Proposition 3.5.4. *Let* L *be a skew field with an automorphism* σ *of order* n *and a primitive root of* 1, ω, *in its centre. Then for any* $a \in L$, $t^n - a$ *is either irreducible over* $L[t;\sigma]$ *or splits into factors of the same degree. In particular, if* n *is prime,* $t^n - a$ *is a product of linear factors or irreducible according as*

$$a = xx^{\sigma}\dots x^{\sigma^{n-1}}$$

has a solution or not.

Proof. Let $p = p(t)$ be an irreducible left factor of $t^n - a$, then so is $p(\omega^{\nu}t)$, for $\nu = 1,\dots,n-1$, therefore $t^n - a = p_1 p_2 \dots p_r q$, where $p_1 = p(t)$ and $p_i = p(\omega^{\nu_i}t)$. If r is chosen as large as possible, each $p(\omega^{\nu}t)$ is a factor of $p_1 p_2 \dots p_r$; in fact this is their least common right multiple and so is unchanged by the substitution $t \longmapsto \omega t$. This means that it is a polynomial in t^n, of positive degree, and a factor of $t^n - a$. Hence it must be $t^n - a$, i.e. $q = 1$ and we have proved the first part.

Now if p has degree d, then $d|n$, hence when n is prime, $d = 1$ or n, and now the last part follows from the identity

$$t^n - a = (t - b)(t^{n-1} + t^{n-2}b^{\sigma^{n-1}} + \dots + b^{\sigma}b^{\sigma^2}\dots b^{\sigma^{n-1}}) + \\ + bb^{\sigma}\dots b^{\sigma^{n-1}} - a. \blacksquare$$

We now turn to the case where n is a power of the characteristic, $n = p^e$, $p = \text{char } K$. Observe that in this case (2) reduces to $D^n = 0$.

Proposition 3.5.5. *Let* L/K *be an outer cyclic extension of degree* $n = p^e$, $p = \text{char } K$, *with automorphism* $\sigma = D + 1$, *Write* $L_{\nu} = \{x \in L \mid x^{(\nu)} = 0\}$, *then each* L_{ν} *is a right* K-*space of dimension* ν *and*

$$K = L_1 \subset L_2 \subset \ldots \subset L_n = L, \qquad L_\nu = L'_{\nu+1} \quad (\nu = 1,\ldots,n-1)$$

Proof. By Th.3.5.1, $[L_\nu:K]_R \leq \nu$ and for $\nu = n$ we have equality. We shall use induction on $n-\nu$, thus assume that $L_{\nu+1}$ has a right K-basis $a_o = 1, a_1, \ldots, a_\nu$. We claim that $a'_1, \ldots, a'_\nu$ are a K-basis for L_ν. If $\Sigma a'_i \alpha_i = 0$, then $\Sigma a_i \alpha_i = \alpha \in K$ and by the linear independence of $a_o, \ldots, a_\nu$ we have $\alpha_1 = \ldots = \alpha_\nu = 0$, thus $a'_1, \ldots, a'_\nu$ are linearly independent; they belong to L_ν so this shows that $L_\nu = L'_{\nu+1}$ and $[L_\nu:K]_R = \nu$. ■

Since $L' = L_{n-1}$, we have

Corollary 1. *The equation* $x' = a$ $(a \in L)$ *has a solution in* L *if and only if* $a \in L_{n-1}$. ■

Corollary 2. *The subspace* L_{p^i} *is the subfield of* L *fixed by* σ^{p^i} $(i = 0,1,\ldots,e)$.

This follows because $D^{p^i} = (\sigma - 1)^{p^i} = \sigma^{p^i} - 1$. ■

Let us define the *trace* of $a \in L$ as $\mathrm{tr}\, a = \Sigma_o^{n-1} a^{\sigma^\nu}$. Then

$$D^{n-1} = (\sigma - 1)^{n-1} = \frac{(\sigma - 1)^n}{(\sigma - 1)} = \frac{\sigma^n - 1}{\sigma - 1} = \Sigma_o^{n-1} \sigma^\nu,$$

hence

$$\text{(6)} \qquad \mathrm{tr}\, a = a^{(n-1)} \quad \text{for any } a \in L.$$

This formula enables us to prove a normal basis theorem:

Theorem 3.5.6. *The outer cyclic extension* L/K *of degree* $n = p^e$ $(p = \mathrm{char}\, K)$ *has a normal basis, and* $a \in L$ *is primitive if and only if* $\mathrm{tr}\, a \neq 0$.

For $\mathrm{tr}\, a \neq 0 \Longleftrightarrow a^{(n-1)} \neq 0 \Longleftrightarrow a \notin L_{n-1}$. Thus for any $a \notin L_{n-1}$, $a^{(\nu)} \in L_{n-\nu}$, $a^{(\nu)} \notin L_{n-\nu-1}$ and hence $a, a', \ldots, a^{(n-1)}$ form a basis of $L_n = L$. ■

We shall only determine extensions of degree p, the case p^e, $e > 1$, follows by repetition (for details see Amitsur

[54]). We shall write $x^{P} = x^{p} - x$; let us also recall the Jacobson-Zassenhaus formula (Jacobson [62],p.187(63)):

$$(x + y)^{p} = x^{p} + y^{p} + \Lambda(x,y),$$

where Λ is a sum of commutators in x and y. It follows that the expression V(x) defined by

$$(7) \qquad V(x) = (t + x)^{P} - t^{P} = (t + x)^{p} - t^{p} - x$$

when evaluated in K[t;1,D] is a polynomial in $x, x', \ldots, x^{(p)}$, since e.g. $[x,t] = xt - tx = x'$. We first prove an analogue of Prop.3.5.4.

Proposition 3.5.7. *Let* L *be a field of characteristic* p *with a derivation* D *such that* $D^{p} = 0$. *For any* $a \in L$, *the polynomial* $t^{P} - a$ *is a product of linear factors over* L[t;1,D] *or irreducible according as the equation*

$$V(x) + a = 0$$

where V *is as in* (7), *has a solution in* L *or not.*

Proof. Let h be a monic irreducible factor of $t^{P} - a$, of degree d say, then the polynomials $h(t + \nu)$ ($\nu = 0,1,\ldots,$ p-1) are factors of $t^{P} - a$. Their least common right multiple is of degree $\geq p$ and is a factor of $t^{P} - a$, hence it must be $t^{P} - a$. Now all the $h(t + \nu)$ are irreducible of the same degree, so $d|p$ and either d = p or d = 1. If $V(b) + a = 0$, then

$$(t + b)^{P} - t = V(b) = -a,$$

hence $t^{P} - a = (t + b)^{P} = (t + b)((t + b)^{p-1} - 1)$ and so $t^{P} - a$ splits into linear factors. Conversely, if $t^{P} - a$ has a linear factor t + b, then $(t + b)^{P} - V(b) - a =$

$(t + b)h(t)$, hence $V(b) + a$ has $t + b$ as factor. But it has degree 0 in t, so $V(b) + a = 0$. ■

We can now prove an analogue of Th.3.5.3.

Theorem 3.5.8. *A skew field* K *of characteristic* p *has an outer cyclic extension of degree* p *if and only if there is a derivation* D *in* K *such that* (i) D^p *is inner, induced by* $a \in K$ *with* $a^D = 0$, *but* D *is outer* (ii) $V(x) + a = 0$ *has no solution in* K.

When this holds, $t^p - a$ is right invariant irreducible in $R = K[t;1,D]$ and $L = R/(t^p - a)R$, with generating automorphism

$$\sigma : \Sigma t^\nu c_\nu \longmapsto \Sigma (t + 1)^\nu c_\nu .$$

Proof. Again (i),(ii) ensure that $t^p - a$ is central and irreducible, so when they are satisfied we have an extension.

Conversely, let L/K be an outer cyclic extension of degree p, then by Prop.3.5.5, L has an element y such that $y^\sigma = y + 1$ Hence $c \longmapsto c^D = cy - yc$ induces a derivation on K and we have a homomorphism $K[t;1,D] \longrightarrow L$ with $t \longmapsto y$. Here $y^p = a \in K$, so D^p is inner, induced by a, and $a^D = 0$, while $V(x) + a = 0$ has no solution in K, by the irreducibility of $y^p - a$ over K. ■

4 · The general embedding

4.1 The category of epic R-fields and specializations

We now come to the fourth method of embedding rings in fields listed in the prologue. It is quite general in that it provides a criterion for arbitrary rings to be so embeddable, and also gives a survey over the different possible embeddings.

Let R be any ring. We shall be interested in R-rings that are fields, R-*fields* for short. If K is an R-field which is generated (as a field) by the image of R, we call K an *epic* R-field. In fact K is an epic R-field precisely when the canonical map $R \longrightarrow K$ is an epimorphism in the category of rings. Our object is to make the epic R-fields (for a given R) into a category and we must find the morphisms. To take R-ring homomorphisms would be too restrictive, for if $f:K \longrightarrow L$ is such a map between epic R-fields, then f is injective (because the kernel is a proper ideal of a field) and im f is a subfield of L containing the image of R, hence im f = L (because L was epic), so f must be an isomorphism. To obtain a workable notion of morphism let us define a *local homomorphism* between any rings A, B as a homomorphism $f:A_o \longrightarrow B$ whose domain A_o is a subring of A and which maps non-units to non-units. If B is a field, this means that the non-units in A_o form an ideal, viz. ker f, hence A_o is then a local ring. Generally by a *local ring* we understand a ring A_o in which the non-units form an ideal $\mathfrak{m}$; the quotient ring $A_o/\mathfrak{m}$ is then a field, called the *residue-*

class field of A_o. Of course when we are dealing with R-rings, a local homomorphism is understood to have a domain which includes the image of R.

Let f be a local homomorphism between epic R-fields K,L. If its domain is K_o, then by what has been said, K_o is a local ring with residue class field $K_o/\ker f$; this is isomorphic to a subfield of L containing the image of R, and hence L. Thus

(1) $K_o/\ker f \cong L.$

Two local homomorphisms are said to be equivalent if their restrictions to the intersection of their domains agree and again define a local homomorphism. This is easily verified to be an equivalence; an equivalence class of local homomorphisms from K to L is also called a *specialization*. It can be checked that the composition of specializations is again a specialization (i.e. composition of mappings, when defined, is compatible with the equivalence defined earlier), and so we obtain for each ring R, a category F_R of epic R-fields and specializations.

At first sight it looks as if there may be several specializations between a given pair of epic R-fields. E.g. let R = k[x,y] be the commutative polynomial ring over a field, K = k(x,y) its field of fractions with the natural embedding and L = k with the homomorphism $R \longrightarrow L$ given by $x \mapsto 0$, $y \mapsto 0$. We obtain a specialization from K to L by defining a homomorphism $\alpha : k[x,y] \longrightarrow L$ in which $x\alpha = y\alpha = 0$. Let K_o be the localization of k[x,y] at the maximal ideal (x,y), then α can be extended in a natural way to K_o. We observe that there are local homomorphisms from K to L that are defined on larger local subrings than K_o (we can 'specialize' rational functions $\phi(x,y)$ so that x/y takes on a specified value in k), but all agree on K_o, so that there is just one specialization from K to L. In fact this is a general property: between any two epic R-fields there is at most one specialization. This will become clear later.

Of course for some rings R there will be no R-fields at all, e.g. R = 0, or for a less trivial example, any simple ring with zero-divisors, say a matrix ring over a field. For any map $R \longrightarrow K$ must be injective and this is impossible when K is a field. Even entire rings R without R-fields exist, e.g. if R is any ring without invariant basis number (Leavitt [57], Cohn [66']); R may be chosen entire and any R-ring is again without invariant basis number and so cannot be a field.

What can we say about R-fields in the commutative case? Let R be a commutative ring and K an epic R-field, then K is of course also commutative (being generated by a homomorphic image of R). The kernel $\mathfrak{p}$ of the natural mapping $R \longrightarrow K$ is a prime ideal and K can be constructed in two ways from R and $\mathfrak{p}$. Firstly we can form $R/\mathfrak{p}$, an integral domain (because $\mathfrak{p}$ is prime), and now K is obtained as the field of fractions of $R/\mathfrak{p}$. Secondly, instead of putting the elements in $\mathfrak{p}$ equal to 0, we can make the elements outside $\mathfrak{p}$ invertible, by forming the localization $R_{\mathfrak{p}}$. This is a local ring and its residue-class field is isomorphic to K. The situation can be illustrated by the accompanying commutative diagram. The two triangles correspond to the two methods of constructing K. The route via the lower triangle is perhaps more familiar,

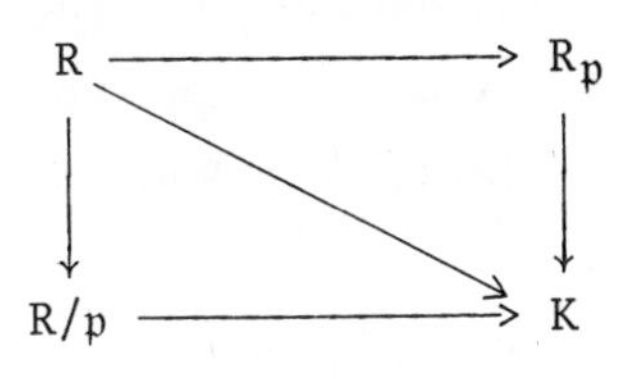

but unfortunately it does not seem to generalize to the non-commutative case; we therefore turn to the upper triangle. Even this cannot be used as it stands, for as we have seen, the field of fractions need not be unique, which means that in general an epic R-field will not be determined by its kernel alone.

Thus to describe an epic R-field we need more than the elements which map to zero, we need the matrices which become singular. Here we use the fact that for any square matrix A over a field K (even skew) the following four conditions are equivalent:

A has no $\genfrac{}{}{0pt}{}{\text{left}}{\text{right}}$ inverse,

A is a $\genfrac{}{}{0pt}{}{\text{left}}{\text{right}}$ zero-divisor.

A matrix A with these properties is called *singular*, all others (if square) are called *non-singular*. Given any R-field $\lambda:R \longrightarrow K$, by the *singular kernel* of K (or λ) written Ker λ, we understand the collection of all square matrices over R(of all orders) which map to singular matrices over K. Let $\mathcal{P}$ be the set of all such matrices, then we can define a localization $R_{\mathcal{P}}$ (analogous to $R_{\mathfrak{p}}$ in the commutative case) as follows. In 1.2 we met the notion of a universal S-inverting ring; we shall need the corresponding construction when S is replaced by a set of matrices over R.

Let Σ be a set of matrices over R, possibly of different orders, but all square (this is to avoid pathologies, because we want to make the matrices in Σ invertible). For every $n \times n$ matrix A in Σ we choose n^2 symbols a'_{ij} which we adjoin to R, with the defining relations (in matrix form)

$$(2) \qquad AA' = A'A = I, \qquad\qquad \text{where } A' = (a'_{ij}).$$

The resulting ring is denoted by R_Σ and called the *universal Σ-inverting ring*. Clearly the natural homomorphism $\lambda:R \longrightarrow R$ is Σ-inverting, in the sense that all matrices in Σ map to invertible matrices over R_Σ (an inverse being provided by (2)), and every Σ-inverting homomorphism $f:R \longrightarrow R'$ can be factored uniquely by λ (the universal mapping property). The proof is the same as for Prop.1.2.1.

We can now describe the construction of an epic R-field in terms of its singular kernel. Let K be an epic R-field, $\mathcal{P}$ its singular kernel and Σ the complement of $\mathcal{P}$ in the set of all square matrices over R. Thus Σ consists of all square matrices over R which become invertible over K. Then the universal Σ-inverting ring R_Σ is a local ring, with residue-class field K. We shall soon see a proof of this fact, but we note that it does not solve our problem yet. For we would like to know when a collection of matrices is a singular kernel, just as we can tell when a collection of elements of R is a prime ideal. In fact we shall be able to characterize singular kernels in much the same way in which kernels of R-fields in the commutative case are characterized as prime ideals.

4.2 The construction of epic R-fields

A basic step in the construction of an R-field is the description of its elements as components of the solution vector of a matrix equation. Given a Σ-inverting homomorphism $f:R \longrightarrow R'$, the set of all entries of matrices $(Af)^{-1}$, where $A \in \Sigma$, is called the *Σ-rational closure* under f of R in R'. It is not hard to give conditions on Σ for this Σ-rational closure to be a ring; in fact they correspond to the condition of being multiplicatively closed in the commutative case for sets of elements. So we define a set Σ of square matrices over a ring R to be *multiplicative* if it includes the 1×1 matrix 1 and for any A, B $\in \Sigma$ we have $\begin{pmatrix} A & C \\ 0 & B \end{pmatrix} \in \Sigma$ for all matrices C of the right size. In any homomorphism $f:R \longrightarrow R'$ the set of *all* matrices inverted over R' is always multiplicative, for 1 is invertible and if A, B are invertible, so is $\begin{pmatrix} A & C \\ 0 & B \end{pmatrix}$, with inverse

$$\begin{pmatrix} A^{-1} & -A^{-1}CB^{-1} \\ 0 & B^{-1} \end{pmatrix}.$$

The characterization of the rational closure, which is at the basis of all further development (Cohn [71"]) stems from the rationality criterion for formal power series due to Schützenberger [62] and Nivat [68].

Theorem 4.2.1. *Let* R *be a ring and* Σ *a multiplicative set of square matrices over* R. *Given a* Σ*-inverting map* $f:R \longrightarrow R'$ *the following conditions on* $x \in R'$ *are equivalent:*

(a) x *lies in the* Σ*-rational closure under* f *of* R *in* R',

(b) x *is a component of the solution of a matrix equation*

(1) $Au + a = 0$, *where* $A \in \Sigma f$, *and* a *is a column over* Rf,

(c) x *is a component of the solution of a matrix equation*

(2) $Au = e$, *where* $A \in \Sigma f$ *and* e *is a column of the identity matrix.*

Moreover, the set of all these elements x *is a subring of* R' *containing* Rf.

Proof. (c) states that u is a column of A^{-1}, hence (c) $\Longleftrightarrow$ (a) and (2) is a special case of (1), so (c) $\Rightarrow$ (b). To prove (b) $\Rightarrow$ (c) we note that if $Au + a = 0$, then

$$\begin{pmatrix} A & a \\ 0 & 1 \end{pmatrix} \begin{pmatrix} u \\ 1 \end{pmatrix} - \begin{pmatrix} 0 \\ 1 \end{pmatrix} = 0,$$

so when (b) holds, each component satisfies an equation of type (2) and so (c) holds.

To prove that the rational closure is a ring containing Rf we use (b): For any $c \in Rf$ we obtain c as solution of

$1.u - c = 0$. Now let u_1, v_1 be the first components of the solutions of $Au + a = 0$, $Bv + b = 0$, then $u_1 - v_1$ is the first component of the solution of

$$\left(\begin{array}{c|c} A & a_1 \quad 0 \\ \hline 0 & B \end{array}\right) \begin{pmatrix} u_1 - v_1 \\ u' \\ v \end{pmatrix} + \begin{pmatrix} a \\ \\ b \end{pmatrix} = 0,$$

where $A = (a_1,\ldots,a_n)$, $u = \begin{pmatrix} u_1 \\ u' \end{pmatrix}$. Further, u_1v_1 is the first component of the solution of

$$\left(\begin{array}{c|c} A & a \quad 0 \\ \hline 0 & B \end{array}\right) \begin{pmatrix} uv_1 \\ \\ v \end{pmatrix} + \begin{pmatrix} 0 \\ \\ b \end{pmatrix} = 0,$$

and the matrices of these systems lie in Σf, because Σ is multiplicative. This shows that the Σ-rational closure contains Rf and admits sums and products, hence it is a ring. ■

This theorem shows that every component of the rational closure can be obtained as some component u_i of the solution of a matrix equation

$$Au = a.$$

Here A is called the *denominator* of u_i and A_i, the matrix obtained by replacing the ith column of A by a, is called the *numerator* of u_i. This usage is justified by Cramer's rule, which states that when R is commutative,

$$u_i = \frac{\det A_i}{\det A}.$$

In the general case we no longer have this formula (because we do not have determinants), but we have the following substitute, still called *Cramer's rule*:

Proposition 4.2.2. *Let* u_i *be the* i*th component of the solution of* $Au = a$, *where* A *is invertible, and let* A_i *be the matrix obtained by replacing the* i*th column of* A *by* a, *then* u_i *is a* $\left\{ \begin{matrix} \textit{left} \\ \textit{right} \end{matrix} \right\} \left\{ \begin{matrix} \textit{zero-divisor} \\ \textit{unit} \end{matrix} \right\}$ *if and only if* A_i *is one (in the matrix ring).*

Proof. Take $i = 1$ for simplicity and write again $u = \begin{pmatrix} u_1 \\ u' \end{pmatrix}$, then $A_1 = (a, a_2, \ldots, a_n) = (Au, Ae_2, \ldots, Ae_n) =$

$A\begin{pmatrix} u_1 & 0 \\ u' & I \end{pmatrix} = A \begin{pmatrix} u_1 & 0 \\ 0 & I \end{pmatrix} \begin{pmatrix} 1 & 0 \\ u' & I \end{pmatrix}$. Thus A_1 is associated (in the full matrix ring R_n) to $\begin{pmatrix} u_1 & 0 \\ 0 & I \end{pmatrix}$ and now the result follows because being a zero-divisor or unit is preserved by multiplying by a unit or bordering with I. ■

Anticipating a definition from 8.1 we may say that u_1 is stably associated to A_1.

As an application let us show how to construct epic R-fields from their singular kernels. Let K be an epic R-field, $\mu : R \longrightarrow K$ the canonical map and $\mathcal{P}$ the singular kernel. Let $\mathcal{P}'$ be the complement of $\mathcal{P}$ in the set of all square matrices over R, then the universal $\mathcal{P}'$-inverting ring, which should be written $R_{\mathcal{P}'}$, is usually written $R_{\mathcal{P}}$ (just as we write $R_{\mathfrak{p}}$ in the commutative case). From the definition of $\mathcal{P}$ it follows that μ can be factored uniquely by $\lambda : R \longrightarrow R_{\mathcal{P}}$ to give a map $\alpha : R_{\mathcal{P}} \longrightarrow K$. We claim that $R_{\mathcal{P}}$ is a local ring with residue class field K. This will follow if we show that every element not in ker α is invertible. For then ker α is the unique maximal ideal of $R_{\mathcal{P}}$ and its residue class field is a subfield of K containing the image of R, hence equal to K, because K was an epic R-field.

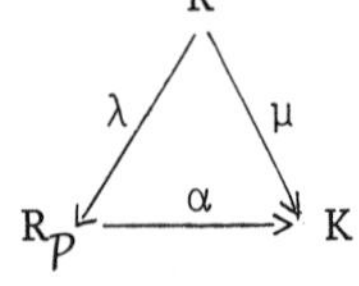

Let $u_1 \in R_{\mathcal{P}}$ be the first component of the solution of an equation $A^{\lambda}u + a = 0$, where A is a matrix over R which becomes invertible over K, and define A_1 as in Cramer's rule. If $u_1^\alpha \neq 0$, then u_1^α is invertible, hence $A_1^{\lambda\alpha} = A_1^\mu$ is invertible, by Cramer's rule, but this means that $A_1 \notin \mathcal{P}$, so A_1^λ is invertible over $R_\mathcal{P}$ and so, again by Cramer's rule, u_1 is a unit in $R_\mathcal{P}$, as claimed.

This result shows that any given epic R-field K can be reconstructed from its singular kernel. What we need now is a simple way of recognizing singular kernels - just as in the commutative case the kernels of epic R-fields are precisely the prime ideals of R. For this we need to develop an analogue of ideal theory in which the place of ideals is taken by certain sets of matrices.

In the first place we must define the operations of addition and multiplication for matrices; they will not be the usual ones of course, but more like the addition and multiplication of determinants.

Multiplication: As the *product* of two square matrices A, B (over any ring R) we take their diagonal sum $A \dotplus B = \begin{pmatrix} A & 0 \\ 0 & B \end{pmatrix}$. Note that over a field $A \dotplus B$ is singular if and only if either A or B is.

Addition is more complicated, just as the addition of determinants is not straightforward, and in fact the latter provides the clue. Let A, B be two matrices which agree in all entries except possibly the first column: $A = (a_1, a_2, \ldots, a_n)$, $B = (a_1', a_2, \ldots, a_n)$, then the *determinantal sum* of A and B is defined as the matrix

$$A \nabla B = (a_1 + a_1', a_2, \ldots, a_n).$$

Similarly one defines determinantal sums with respect to another column or with respect to a row. Of course it must be borne in mind that the determinantal sum need not be defined. As notation we shall always use $A \nabla B$, indicating in words the relevant column or row, when this is necessary to prevent confusion.

We observe that over a commutative ring, where determinants are defined, one has $\det(A \nabla B) = \det A + \det B$, whenever the determinantal sum (for any row or column) is defined. Likewise, over a skew field, if two of A, B, $A \nabla B$ are singular, so is the third. Over a general ring there is no direct interpretation, but, and this is the point, whether the operation is defined depends only on the matrices involved and not on the ring.

The third operation we need is the analogue of zero, in our case a matrix which becomes singular under any homomorphism into a field. Here one cannot just take zero-divisors, for if $AB = 0$, where $A,B \neq 0$, it may still happen that under a homomorphism A becomes invertible and B becomes zero. But there are some matrices that always map to singular ones, e.g. the zero matrix, and more generally a matrix of the form $\begin{pmatrix} ac & ad \\ bc & bd \end{pmatrix} = \begin{pmatrix} a \\ b \end{pmatrix}(c\ d)$. Let us call an $n \times n$ matrix A *non-full* if $A = PQ$, where P is $n \times r$, Q is $r \times n$ and $r < n$; in the contrary case A is *full*. For example, a 1×1 matrix is full if and only if it is non-zero.

We can now define the notion of a *matrix ideal*. By this we understand, in any ring R, a collection $\mathcal{A}$ of square matrices satisfying the following four conditions:

1. $\mathcal{A}$ contains all non-full matrices,
2. If $A, B \in \mathcal{A}$, then $A \nabla B \in \mathcal{A}$ whenever this is defined,
3. If $A \in \mathcal{A}$, then $A \dotplus B \in \mathcal{A}$ for all square matrices B,
4. If $A \dotplus 1 \in \mathcal{A}$, then $A \in \mathcal{A}$.

If moreover,

5. $\mathcal{A}$ is proper, i.e. it does not include all square matrices over R,
6. $A, B \notin \mathcal{A} \Rightarrow A \dotplus B \notin \mathcal{A}$,

then $\mathcal{A}$ is called a *prime matrix ideal*. The analogy with prime ideals is clear, at least at the formal level. We now quote (without proof) results which show that the analogy is more than formal.

Theorem 4.A. *Let* Σ *be any set of square matrices over a ring* R, *closed under diagonal sums and containing* 1, *and let* $\mathcal{A}$ *be a matrix ideal disjoint from* Σ, *then* (i) *there exist maximal matrix ideals containing* $\mathcal{A}$ *and disjoint from* Σ *and* (ii) *any such matrix ideal is prime.*

Of course this is the precise analogue of what happens in the commutative case and the proof is very similar, also using Zorn's lemma (cf. Cohn [71"],ch.7). For any matrix ideal $\mathcal{A}$ one can define the radical as

$$\sqrt{\mathcal{A}} = \{A \mid \underbrace{A \dotplus \dots \dotplus A}_{r} \in \mathcal{A}, \text{ for some } r\}.$$

It is easily checked that $\sqrt{\mathcal{A}}$ is a matrix ideal containing $\mathcal{A}$, and in fact, using Th.4.A, one proves easily that

$$\sqrt{\mathcal{A}} = \cap \{\mathcal{P} \mid \mathcal{P} \text{ prime} \supseteq \mathcal{A}\},$$

again the analogue of a well known result in commutative algebra. It is not difficult to verify that the singular kernel of an epic R-field is a prime matrix ideal. The basic result in what follows is the converse:

Theorem 4.B. *Let* R *be any ring, then the singular kernel of any epic* R*-field is a prime matrix ideal and conversely, any prime matrix ideal of* R *is the singular kernel of some epic* R*-field.*

We already know how to construct the epic R-field defined by a given prime matrix ideal $\mathcal{P}$: it is the residue class field of the localization $R_{\mathcal{P}}$. The proof of Th.4.B, though long, is not difficult: it constructs the epic R-field as a set of equivalence classes of solutions of matrix equations.

Th.4.B is the key result which enables us to answer the questions posed earlier. Thus R has an R-field at all precisely if it has a prime matrix ideal, and this is so if and only if the least matrix ideal (generated by the empty set) is proper. If one transforms this condition slightly, one finds

Corollary 1. *A ring* R *has a homomorphism into a field if and only if the unit matrix* I *(of any size) cannot be written as a determinantal sum of non-full matrices.*

More explicitly, if R has no R-field, then there must be an equation

$$(3) \qquad I = C_1 \nabla C_2 \nabla \ldots \nabla C_r,$$

where the C_i are non-full, and the right-hand side is bracketed in some way so as to make sense. This is a very explicit condition, but if we are given a ring R with no R-fields it may be far from easy to construct an equation (3); the proof will be of no help since it was non-constructive (Zorn's lemma was used).

The situation may be illustrated by a corresponding problem for commutative rings. If R is a commutative integral domain, then any n x n matrix A which is nilpotent satisfies $A^n = 0$, as we see by transforming A to triangular form over the field of fractions of R. It follows that for any commutative ring R, if A is an n x n nilpotent matrix, then the entries of A^n must lie in every prime ideal of R and hence be nilpotent. In particular, let A be a 2 x 2 matrix over a commutative ring R and denote by J the ideal generated by the entries of A^3, then the entries of A^2 lie in $\sqrt{J}$. The explicit verification shows that such a problem is not always trivial (if $A^2 = \begin{pmatrix} \alpha & \beta \\ \gamma & \delta \end{pmatrix}$, it is found that $\beta^9, \gamma^9, \alpha^{11}$, δ^{11} lie in J).

Returning to Cor.1, to give an example, consider a ring

R with an $r \times s$ matrix A and an $s \times r$ matrix B such that $AB = I$. If $s < r$, there are no R-fields; this follows trivially (without Cor.1) because the $r \times r$ unit matrix is not full. What Cor.1 shows is that a slightly more general condition is sufficient as well as necessary for the existence of R-fields.

To get a field of fractions we need an injective homomorphism, i.e. the singular kernel must not contain any non-zero elements of R. Let Σ be the set of all diagonal matrices with non-zero entries on the main diagonal; we need a prime matrix ideal disjoint from Σ and by Th.4.A this exists if and only if the least matrix ideal does not meet Σ. This gives

Corollary 2. *A ring* R *has a field of fractions if and only if no diagonal matrix with non-zero diagonal elements can be expressed as a determinantal sum of non-full matrices.*

Here the absence of zero-divisors, though clearly necessary, is not postulated explicitly. In fact it follows from the given condition, for if $ab = 0$, then

$$\begin{pmatrix} a & 0 \\ 0 & b \end{pmatrix} = \begin{pmatrix} a & 0 \\ 1 & b \end{pmatrix} \nabla \begin{pmatrix} 0 & 0 \\ -1 & b \end{pmatrix} = \begin{pmatrix} a \\ 1 \end{pmatrix} (1 \quad b) \nabla \begin{pmatrix} 0 \\ 1 \end{pmatrix} (-1 \quad b) .$$

Although this solves our problem we now see that we would like to know more: Is there more than one field of fractions, and if so, is there one that is universal in some sense? Clearly by a *universal* R-*field* one understands an epic R-field which has every other epic R-field as specialization. If K_1, K_2 are epic R-fields with singular kernels P_1, P_2 respectively, then there is a specialization from K_1 to K_2 if and only if $P_1 \subseteq P_2$. This means that the category of epic R-fields and specializations is equivalent to the set of prime matrix ideals, regarded as a category by its partial ordering by inclusion. It shows that there is a universal R-field precisely if there is a least prime matrix ideal.

We have presented these results without proofs because these are long (and can be found in Cohn [71"]ch.7); moreover, a knowledge of the proofs is not necessary to apply the results. In fact the general criteria obtained here are quite difficult to apply; this is mainly because calculations with determinantal sums are rather cumbersome. But, to give a comparison, the criterion for a ring to be entire is as simple as one could wish (absence of zero-divisors) and yet it may be very difficult to decide whether a given ring is entire, e.g. it is not yet known whether the group ring of every torsion free group is entire. But there is one important case in which the above criteria can be used to prove the existence of a universal field of fractions, and we turn now to describe this case.

4.3 Firs and semifirs

For any ring R we can form the least matrix ideal N and we have seen that (i) R has an R-field if and only if N is proper, (ii) R has a universal R-field if and only if $\sqrt{N}$ is prime, while generally there will be more R-fields, the smaller N is. At the very least N will have to contain all non-full matrices, so the most favourable situation is that where N consists precisely of all non-full matrices. This is described in

Theorem 4.3.1. *A ring* R *has a universal field of fractions* **F** (**R**) *over which every full matrix can be inverted if and only if*

(i) $1 \neq 0$ *and the diagonal sum of any full matrices is full,*

(ii) *the determinantal sum of any non-full matrices, whenever defined, is non-full.*

For (ii) ensures that the non-full matrices form a matrix ideal, which must be prime, by (i). ■

For commutative rings (i) ensures that R mod its nilradical is an integral domain. To give an example of a ring satisfying (i) but not (ii), we recall that a ring is said to be *weakly finite* if for any two n x n matrices A, B, AB = I implies BA = I (all $n \geq 1$). A weakly finite ring has invariant basis number and the unit matrix of any size is full. Now we have

Proposition 4.3.2 *(Klein* [69]*). Any ring satisfying Klein's nilpotence condition is weakly finite.*

Proof. Let $A, B \in R_n$ satisfy $AB = I$, then $A^r B^r = I$ for all $r \geq 1$. Clearly $A(I - B^r A^r) = (I - B^{r-1} A^{r-1})A$, hence the matrix $C = A(I - B^r A^r)$ satisfies $C^i = (I - B^{r-i} A^{r-i})A^i$, in particular, $C^r = 0$ and choosing $r = n+1$ we find that $I - BA = (I - BA)A^n B^n = C^n B^n = 0$. ■

Consider the following example due to Bergman [74]. Let R be defined by 27 elements, arranged as 3 x 3 matrices P, U, V with defining relations $UV = VU = I$, $P^2 = P$, $UP = (I - P)U$. If R could be mapped into a field, then P could be transformed to diagonal form, with 0's and 1's on the main diagonal; since P is similar to I - P, there must be equal numbers of both, but this is impossible, because the order is odd. It is easy to see that R is entire and Bergman shows that it satisfies Klein's nilpotence condition, therefore it is weakly finite, in particular I is full. But by Cor.1 of Th.4.B, the unit matrix (of a certain size) can be written as a determinantal sum of non-full matrices. Thus R does not satisfy (ii).

The conditions of Th.4.3.1 are not easy to apply; there is just one case where they can be checked without difficulty, namely for semifirs, which were introduced in 1.1: Once the basic properties of semifirs have been derived, such a verification takes less than a page (Cohn [71"], p.283), but since we have not developed this background here, we omit a detailed proof:

Theorem 4.C. *Every semifir has a universal field of fractions, obtained as the universal ring inverting all the full matrices.*

We shall denote the universal field of fractions of R by **F**(R). To prove the result one only has to verify (i),(ii) of Th.4.3.1. Here (i) follows from a form of Sylvester's law of nullity for semifirs, while (ii) is a relatively direct calculation, based on the dimension formula

$$\dim(U + V) + \dim (U \cap V) = \dim U + \dim V$$

for finitely generated submodules of free modules over a semifir.

To apply Th.4.C we derive a further consequence which tells us when a homomorphism can be extended. A homomorphism $f:R \longrightarrow S$ between any rings is called *honest* if it keeps full matrices full. Any homomorphism keeps non-full matrices non-full, hence any isomorphism and in particular any automorphism is honest. An honest homomorphism must be injective, for an element c is non-zero if and only if it is full, as 1 x 1 matrix. But an injective homomorphism need not be full; here is an example.

Let $R = k\langle x_1,x_2,y_1,y_2\rangle$ be a free k-algebra on four generators and define an endomorphism α over k by $x_i \longmapsto x_1y_i$, $y_i \longmapsto x_2y_i$ $(i = 1,2)$. It is easily checked that α is injective; but it is not honest, as the equation

$$\begin{pmatrix} x_1 & x_2 \\ y_1 & y_2 \end{pmatrix}^{\alpha} = \begin{pmatrix} x_1y_1 & x_1y_2 \\ x_2y_1 & x_2y_2 \end{pmatrix} = \begin{pmatrix} x_1 \\ x_2 \end{pmatrix} (y_1 \quad y_2)$$

shows. The right-hand side is not full, but the matrix to which α is applied is full, since it can be specialized to the unit-matrix which is full.

Theorem 4.3.3. Let $f:R \longrightarrow S$ *be a homomorphism between semifirs. Then* f *extends to a homomorphism (necessarily unique) between their universal fields of fractions if and only if it is honest. In particular, every isomorphism between* R *and* S *extends to a unique isomorphism between their universal fields of fractions.*

Proof. Denote by Φ, Ψ the set of all full matrices over R, S respectively, then if f is honest, $\Phi^f \subseteq \Psi$, and so the mapping $R \longrightarrow S \longrightarrow S_\Psi$ is Φ-inverting. Hence there is a unique homomorphism $f_1:R_\Phi \longrightarrow S_\Psi$ such that the diagram shown commutes, i.e. f can be extended (in just one way). Conversely, if an extension of f exists, any full matrix A over R becomes invertible over R_Φ and is mapped to an invertible matrix over S_Ψ. But this is the image of A^f, which must therefore be full. Hence f is honest, as claimed. The rest follows since an isomorphism is always honest. ■

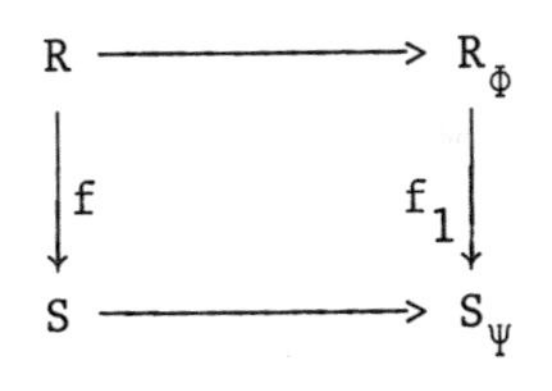

The notion of an 'honest map' is chiefly of use for semifirs, because here the non-full matrices constitute the unique least prime matrix ideal.

To get an idea of the usefulness of Th.4.C and Th.4.3.3 we really need to know how extensive the class of semifirs is. In the commutative case semifirs are just Bezout domains. Somewhat more familiar are the principal ideal domains; they may also be characterized as the Noetherian Bezout domains. Analogously the semifirs contain as subclass the *firs* (= free ideal rings), which by definition are rings in which every right ideal and every left ideal is free of unique rank. This class is far more extensive than the class of non-commutative principal ideal domains. To give some examples, firs include (i) free algebras over a field, (ii) group algebras of free groups, (iii) free products of skew fields. These examples will be examined in more detail in Ch.5. For the moment we only note that for any field k and any set X, the free algebra $k<X>$ is a fir; this is most easily proved by the weak algorithm, a form of the Euclidean algorithm adapted for use in free algebras (cf. Cohn [71''],Ch.2).

Earlier in 1.1 we quoted the result from (Cohn [69]) showing the existence of n-firs from (n+1)-term relations without interference. This can be used to show that for any $n \geq 1$ there exists an n-fir not embeddable in a field. We take $2(n+1)^2$ elements a_{ij}, b_{ij} $(i,j = 1,\ldots,n+1)$ arranged as matrices $A = (a_{ij})$, $B = (b_{ij})$ with relations (in matrix form) $AB = I$. These relations satisfy the non-interference condition, so we have an n-fir, but $BA \neq I$ (by an easy normal form argument), so the ring constructed is not weakly finite and therefore not embeddable in a field. It we take $2(n+1)(n+2)$ variables $a_{i\lambda}, b_{\lambda i}$ $(i = 1,\ldots, n+1,\ \lambda = 1,\ldots, n+2)$ with relations $AB = I_{n+1}$, $BA = I_{n+2}$, we get an n-fir with no R-field.

These rings are of interest in that they enable us to answer the following problem raised by Mal'cev [73]. As we saw in 1.1, the class of entire rings embeddable in fields can be defined by quasi-identities, and we note incidentally that the conditions implicit in Cor.1,2 of Th.4.2.3 are easily put into this form. Now Mal'cev asks whether this class can already be defined by a *finite* set of quasi-identities. The answer is 'no' (as for semigroups) and this may be seen as follows: Suppose there is a finite set of first-order sentences which expresses the fact that a ring is embeddable in a field. On replacing them by their conjunction we obtain a single sentence A say, which is necessary and sufficient for a ring to be embeddable in a field. Now let $\mathcal{F}_n$ be the class of n-firs for which A is false, then F_n is an elementary class (cf. e.g. Cohn [65]) since n-firs can be defined by an elementary sentence. Now $\cap F_n$ is the class of semifirs not satisfying A, but every semifir is embeddable in a field and so satisfies A, hence $\cap F_n = \emptyset$. Thus we have a family of elementary model classes with empty intersection; by the compactness theorem in logic (cf. e.g. Cohn [65], Mal'cev [73]) $F_n = \emptyset$ for some n, i.e. there exists n such that every n-fir satisfies A and so is embeddable in a field. But this contradicts our earlier findings and we conclude (Cohn [74"]):

Theorem 4.3.4. *The condition for a ring to be embeddable in a field cannot be expressed in a finite set of sentences.*

In intuitive (though imprecise) terms we can say that embeddability in a field requires n-term conditions for arbitrarily large n. This is in interesting contrast with embeddability in a group. If R is entire, so that R* is a semigroup, a sufficient condition for the embeddability of R* in a group can be expressed in terms of 2-term conditions, for we have the following result (Cohn [71]):

Theorem 4.D. *Let* R *be a* 2-*fir in which every non-unit is a finite product of irreducibles (i.e.* R *is 'atomic'), then* R* *is embeddable in a group.*

This makes the difference between embeddability in a group and in a field rather clear, and it provides a simple answer to another question of Mal'cev's, whether an entire ring R exists such that R* is embeddable in a group but R is not embeddable in a field. To get an example we need only take 24 generators $A = (a_{i\lambda})$, $B = (b_{\lambda i})$ $(i = 1,2,3,$ $\lambda = 1,2,3,4)$, such that $AB = I_3$, $BA = I_4$. This is an atomic 2-fir (Cohn [69]), hence R* is embeddable in a group, but there are no R-fields. Other examples, using similar principles, were found by A.J. Bowtell [67] and A.A. Klein [67]. L.A. Bokut [69] also gives an example of such a ring; his construction is more complicated, but unlike the other cases, his example is of a semigroup algebra.

5 · Coproducts of fields

5.1 The coproduct construction for groups and rings

Let $\mathcal{A}$ be any category; we recall the definition of the coproduct. Given any family (A_i) of objects in $\mathcal{A}$, and an object S with a family of maps $\mu_i : A_i \longrightarrow S$, this defines a natural correspondence $f \longmapsto \mu_i f$ from maps $S \longrightarrow X$ to families of maps $A_i \longrightarrow X$, thus a mapping

$$\mathcal{A}(S,X) \longrightarrow \prod \mathcal{A}(A_i,X).$$

When this mapping is a bijection, the object S with the maps μ_i is called the *coproduct* of the A_i and is written $\coprod A_i$. From the definition it is easily seen to be unique up to isomorphism, if it exists at all. Thus for sets we obtain the disjoint union, for abelian groups the direct sum, for general groups the free product, but we shall return to this case below.

Often we need an elaboration of this idea. Let K be a fixed object in $\mathcal{A}$ and consider the *comma category* $(K,\mathcal{A})$: its objects are arrows $K \longrightarrow A$ ($A \in \text{Ob}\,\mathcal{A}$) and its morphisms are commutative triangles

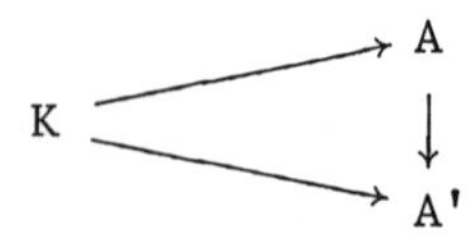

This category has the initial object $K \xrightarrow{1} K$; it reduces

to A when K is an initial object of A. Now the coproduct in (K,A) is called the *coproduct over* K. E.g., for two objects $K \longrightarrow A$, $K \longrightarrow B$, this is just their pushout.

Consider coproducts over a fixed group K in the category of groups. This means that we have a family of groups (G_i) and homomorphisms $\phi_i : K \longrightarrow G_i$ and the coproduct C with maps $\alpha_i : G_i \longrightarrow C$ is a sort of 'general pushout'.

Clearly any element of K mapped to 1 by any ϕ_i must be mapped to 1 by every α_j, so by modifying K and the G_i we may as well assume that each ϕ_i is injective; this means that K is embedded in G_i via ϕ_i. If in this situation all the α_i are injective, the coproduct is called *faithful*. If moreover, $G_i\alpha_i \cap G_j\alpha_j = K\phi_i$ for all $i \neq j$, the coproduct is called *separating*. These definitions apply quite generally for concrete categories (i.e. categories where the objects have an underlying set structure). Now for groups we have the following basic result (Schreier [27]):

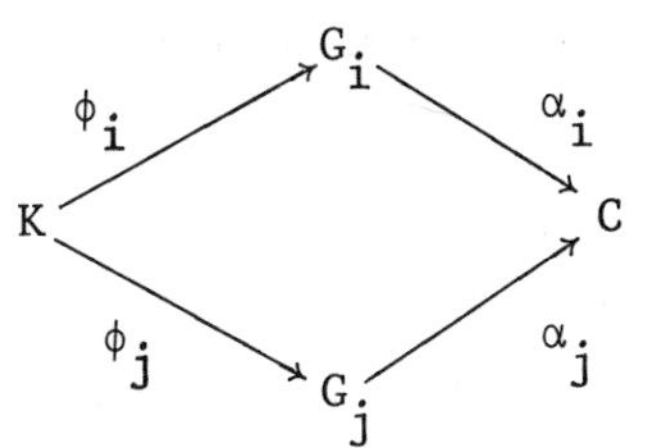

Theorem 5.1.1. *The coproduct of groups (over a fixed group) is faithful and separating.*

This is proved by writing down a normal form for the elements of the coproduct C. Let $K_i = K\phi_i$ be the image of K in G_i and choose a left transversal for K_i in G_i of the form $S_i \cup \{1\}$, thus $G_i = K_i \cup S_iK_i$. Then every element of C can be written in just one way as

$$(1) \quad u_1u_2\ldots u_nc \quad (n \geq 0;\ u_\nu \in S_{i_\nu},\ i_{\nu-1} \neq i_\nu,\ c \in K).$$

It is clear how to write any element of C in this form; to prove the uniqueness one defines a multiplication of the expressions (1) (which consists in a set of rules reducing the formal product of two expressions to normal form), and verifies that one obtains a group in this way. A quicker way (v.d. Waerden [48]) is to define a group action on the set of expressions (1) for each group G_i:

$$u_1 \dots u_n c.g = \begin{cases} u_1 \dots u_n u_p c' & \text{if } i_n \neq i \text{ and } cg = u_p c' \text{ in } G_i, \\ u_1 \dots u_{n-1} u_p c' & \text{if } i_n = i \text{ and } u_n cg = u_p c' \text{ in } G_i, \end{cases}$$

with the understanding that u_p is omitted if $cg \in K_i$ (or in the second case, if $u_n cg \in K_i$). Now it can be verified that these group actions can be combined to give a C-action on the elements (1), and the conclusion follows because the expressions (1) are distinct.■

We have given this proof in outline since it is very similar to the corresponding proof for the coproduct of fields, which we shall soon meet.

The coproduct of groups over a given group K is usually called the *free product of groups with amalgamated subgroup* K. It is also possible to define a notion of free product of groups with different amalgamated subgroups, where we have a family of groups (G_i) and subgroups H_{ij} of G_i such that $H_{ij} \cong H_{ji}$. This can again be constructed as a coproduct, amalgamating H_{ij} with H_{ji}, but it will in general be neither faithful nor separating (cf. B.H. Neumann [54] and the references given there).

Our aim in this section is to describe the coproduct of rings. Thus let K be a fixed ring and consider K-rings; it is easy to see that coproducts always exist. We simply take a presentation by generators and defining relations for each K-ring in our family and write all these presentations together. But the coproduct need not be faithful or separating. Before finding conditions for it to be so we look at some examples.

1. Let k be a commutative field, $K = k[\lambda]$, where λ is a central indeterminate, $R = k(\lambda)$, $S = k[\lambda,\mu \mid \lambda\mu = 0]$. In $R \underset{K}{\sqcup} S$, the coproduct of R and S over K, we have $\mu = 1.\mu = \lambda^{-1}.\lambda\mu = 0$, so S is not faithfully represented.
2. The inclusion $\mathbf{Z} \subset \mathbf{Q}$ is an epimorphism, and $\mathbf{Q} \underset{\mathbf{Z}}{\sqcup} \mathbf{Q} = \mathbf{Q}$. Hence the coproduct is faithful, but not separating. More generally, if $R \longrightarrow S$ is any ring epimorphism, one finds that $S \underset{R}{\sqcup} S \cong S$ (cf. Knight [70]).

When the coproduct of rings is faithful and separating we shall often call it the *free product*. Thus the question is: when does the free product exist? We begin by giving a necessary condition; for simplicity we limit ourselves to two factors.

Let R_1, R_2 be K-rings which are *faithful*, i.e. the mapping $K \longrightarrow R_i$ is injective. If the free product is to exist, we must have

$$c_1a,\ ac_2 \in K \Rightarrow\ c_1a.c_2 = c_1.ac_2 \text{ in } K, \text{ for all } c_i \in R_i,\ a \in K.$$

Clearly this also holds with R_1 and R_2 interchanged, and more generally, for matrix equations. If we take *all* implications of a suitable form we can obtain necessary and sufficient conditions for free products to exist, but they will not be in a very explicit form. Syntactical criteria of a rather different form have been obtained by D.A. Bryars.

We now come to a simple sufficient condition for the existence of free products which will be enough for all the applications we have in mind. This states essentially that the free product of a family $\{R_\lambda\}$ of faithful K-rings exists provided that each quotient module R_λ/K is free as right K-module. A direct proof, following the pattern of the proof of Schreier's theorem 5.1.1, is quite straightforward. However, we shall want to prove a little more and therefore follow a slightly different approach, which will place us in a position to obtain Bergman's coproduct theorem in the next section. I am indebted to W. Dicks for the presentation of these results.

We shall want to consider the coproduct of a quite arbitrary family $\{R_\lambda\}$ $(\lambda \in \Lambda)$ of K-rings. It will be convenient to write $K = R_o$ where we assume that $0 \notin \Lambda$, and to write $\lambda, \lambda', \lambda_o$ etc. for the typical element of Λ and μ, μ', etc. for the typical element of $\Lambda \cup \{0\}$.

Theorem 5.1.2. *Let* R_o *be any ring,* $\{R_\lambda | \lambda \in \Lambda\}$ *a family of faithful* R_o*-rings and* $R = \coprod_{R_o} R_\lambda$ *their coproduct. If the quotient modules* R_λ/R_o *are free as right* R_o*-modules, then the coproduct is faithful and separating, and for each* $\mu \in \Lambda \cup \{0\}$, R/R_μ *(hence also* R *itself) is free as right* R_μ*-module.*

Further, for each $\mu \in \Lambda \cup \{0\}$, *let* M_μ *be a free right* R_μ*-module and put* $M = \bigoplus_\mu M_\mu \otimes_{R_\mu} R$. *Then the canonical* R_μ*-module homomorphism* $M_\mu \longrightarrow M$ *is injective and the cokernel is free as right* R_μ*-module. (Equivalently:* $M = M_\mu \oplus M'_\mu$, *where* M'_μ *is free as right* R_μ*-module).*

Proof. For each $\lambda \in \Lambda$ let $T_\lambda \cup \{1\}$ be a right R_o-basis of R_λ and for each $\mu \in \Lambda \cup \{0\}$ let S_μ be a right R_μ-basis of M_μ. We shall denote the disjoint union of the T_λ by T, that of the S_μ by S and call the elements of T_λ or S_λ *associated with* the index λ; the elements of S_o are not associated with any index.

We begin by proving that $M = \oplus M_\mu \otimes R$ has a right R_o-basis consisting of all products

$$(2) \qquad u = st_1 \ldots t_n \qquad s \in S,\ t_i \in T,\ n \geq 0,$$

where no two successive terms are associated with the same λ. To establish this fact, denote the set of all such formal products by U and let F be the free right R_o-module on U. We shall define a right R-module structure on F and show that $F \cong M$.

An element $u \in U$ is said to be *associated with* λ if its last factor (an element of S or T) is associated with λ. The set of elements of U *not* associated with λ is denoted by U^λ. Fix $\lambda \in \Lambda$, then we may write $F = S_\lambda R_o \oplus (U\backslash S_\lambda)R_o$, and we can give the first summand a right R_λ-structure by identifying it with M_λ; we note that this still holds if we replace λ by 0. Now consider the free right R_λ-module on U^λ as basis: $U^\lambda R_\lambda$. We have an R_o-linear map: $U^\lambda R_\lambda \longrightarrow (U\backslash S_\lambda)R_o$ given by formal multiplication $(u,t) \longmapsto ut$ ($u \in U^\lambda$, $t \in T_\lambda \cup \{1\}$). This map is easily seen to be bijective, so we have an R_λ-module structure on $(U\backslash S_\lambda)R_o$ and in this way F becomes an R_λ-module, for each $\lambda \in \Lambda$; all the R_o-actions agree, so F is in fact an R-module.

Moreover, for each $\mu \in \Lambda \cup \{0\}$ there is an R_μ-linear map $M_\mu \longrightarrow F$ which is injective, with a free complement, so there is an R-linear map $M \longrightarrow F$. To show that this is an isomorphism we construct its inverse. Let $f_\mu : M_\mu \longrightarrow M$ be the canonical R_μ-linear map and define $f : U \longrightarrow M$ by

$$(st_1 \ldots t_n)f = sf_\mu . t_1 \ldots t_n \qquad \text{if } s \in S_\mu .$$

By R_o-linearity this extends to a map $f : F \longrightarrow M$ which is clearly the desired inverse. This proves the second part, and applied to $M_\lambda = R_\lambda$ it shows that the coproduct is faithful. To show that it is separating we take $M_o = R_o$, $M_\lambda = 0$ ($\lambda \in \Lambda$), then we find that $M = R$ has an R_o-basis consisting of all products $t_1 t_2 \ldots t_n$ ($t_i \in T$), where no two successive terms are associated with the same λ. Thus if $1,2 \in \Lambda$, $1 \neq 2$, then the R_o-submodule $R_1 + R_2$ has as basis $T_1 \cup T_2 \cup \{1\}$, hence $R_1 \cap R_2 = R_o$ and this shows that the coproduct is separating. ■

When R_o is a field (the case of main importance for us), all the conditions are easily satisfied and we obtain the

Corollary. The free product of any family $\{R_\lambda\}$ *of non-zero rings over a skew field exists and is left and right free over each* R_λ. ■

The free product is known to exist under more general conditions; instead of requiring R_λ/R_o to be free it is enough for it to be locally free (i.e. every finite subset is contained in a free submodule) or more generally, flat, i.e. each R_λ is faithfully flat (cf. Cohn [59]). But we observe that over a semifir, 'flat' means the same as 'locally free' (Cohn [71"], Th.1.4.4,p.56).

Th.5.1.2 provides a means of finding the homological dimension of certain R-modules, for a coproduct R:

Proposition 5.1.3. *(Bergman [74]). Let* R_o *be a skew field and* $M = \oplus M_\mu \otimes_{R_\mu} R$, *then*

$$\mathrm{hd}_R\, M = \sup_\mu(\mathrm{hd}_{R_\mu} M_\mu).$$

Proof. Clearly it suffices to show that $\mathrm{hd}_R(M_\mu \otimes R) = \mathrm{hd}_{R_\mu} M_\mu$. Now R is left free, hence left flat over R_μ, therefore $- \otimes_{R_\mu} R$ converts a projective R_μ-resolution of M_μ to a projective R-resolution of $M_\mu \otimes R$ and so $\mathrm{hd}_R M_\mu \otimes R \leq \mathrm{hd}_{R_\mu} M_\mu$. To show that equality holds we observe that since R is right R_μ-projective, any R-projective resolution of $M_\mu \otimes R$ is also R_μ-projective, hence $\mathrm{hd}_R\, M_\mu \otimes R \geq \mathrm{hd}_{R_\mu} M_\mu \otimes R$, but the latter equals $\mathrm{hd}_{R_\mu} M_\mu$, because M_μ and $M_\mu \otimes R$ differ only by a free summand. ■

5.2 Projective modules over coproducts over skew fields

In this rather technical section we show that the projective modules of a coproduct $R = \sqcup_{R_o} R_\lambda$ over a skew field

R_o are of the form $\oplus P_\mu \otimes_{R_\mu} R$, where P_μ is a projective R_μ-module. The results are due to Bergman [74], the presentation is again due to Dicks. We fix a skew field R_o and retain the notation of the proof of Th.5.1.2.

For any $\lambda \in \Lambda$ we obtained a direct sum representation $M = M_\lambda \oplus U^\lambda R_\lambda$. Thus for any $u \in U^\lambda$ we have a mapping $p_{\lambda u}$: $M \longrightarrow R_\lambda$, namely the coordinate in R_λ corresponding to the term u. Similarly, since $M = UR_o$, we have for any $u \in U$ a mapping $p_{ou}: M \longrightarrow R_o$. For convenience we write U^o for U, and for any $\mu \in \Lambda \cup \{0\}$ we define the μ-*support* of a subset X of M to be $\{u \in U^\mu | xp_{\mu u} \neq 0$ for some $x \in X\}$. The 0-support will also be called the *support*.

The elements of the basis U will be called *monomials*; the *degree* of a monomial is its length. Let us well-order the sets S and T arbitrarily and then well-order U by degree and monomials of the same degree lexicographically, reading from left to right.

Next well-order $\Lambda \cup \{0\}$, making 0 the least element, and then well-order $(\Lambda \cup \{0\}) \times U$ first by the degree of the second factor, then (for a given degree) lexicographically from left to right. Finally let H be the set of almost-everywhere zero functions $(\Lambda \cup \{0\}) \times U \longrightarrow \mathbf{N}$, well-ordered lexicographically reading from highest to lowest in $(\Lambda \cup \{0\}) \times U$.

Consider any non-zero element x in M; the monomials in the support of x will be called its *terms*. The greatest such monomial (in the ordering of U) is called the *leading term* and its degree is the *degree* of x, written deg x. If all terms of x of degree deg x are associated with λ, x is called λ-*pure*. If x is not λ-pure, the greatest element in the support belonging to U^λ is called the λ-*leading term* (this should perhaps be called the non-λ leading term). If x is λ-pure for some λ, it is called *pure*, otherwise it is *impure* or also 0-*pure*, and its leading term is then called the 0-

leading term.

With these preparations we can state the main result of this section.

Theorem 5.2.1 *(Bergman* [74]). *Let* $R = \coprod_{R_o} R_\lambda$ *be a coproduct, where* R_o *is a skew field, and for any family* $\{M_\lambda\}$, *where* M_λ *is a free* R_λ*-module, put* $M = \oplus M_\lambda \otimes_{R_\lambda} R$. *If* L *is any submodule of* M, *then for each* $\mu \in \Lambda \cup \{0\}$ *there is an* R_μ*-submodule* L_μ *of* L *such that the canonical map* $\oplus L_\mu \otimes_{R_\mu} R \longrightarrow L$ *is an isomorphism.*

Proof. For each $\lambda \in \Lambda$ let L_λ be the R_λ-submodule of L consisting of all elements whose λ-support does not contain the λ-leading term of any (non λ-pure) element of L. By L_o we denote the R_o-module consisting of all elements of L whose support does not contain the leading term of any pure element of L. We claim that the family $\{L_\mu\}$ has the desired properties.

To prove that $\Sigma L_\mu R = L$, assume that this is not so and choose $y \in L$, $y \notin \Sigma L_\mu R$ so as to minimize $h(y) \in H$, where

$$h(y)(\mu,u) = \begin{cases} 1 \text{ if } u \text{ is in the } \mu\text{-support of } y \text{ and } y \text{ is } \mu\text{-pure,} \\ 0 \text{ otherwise.} \end{cases}$$

Suppose first that y is pure, say λ-pure. Since $y \notin L_\lambda$, some monomial u in the λ-support of y is the λ-leading term of some non λ-pure element x of L. Clearly deg x < deg y, therefore $x \in \Sigma L_\mu R$; further, there exists $c \in R_\lambda$ such that the λ-support of y - xc does not contain u, and y - xc is either λ-pure with less λ-support or of lower degree than y, hence $h(y - xc) < h(y)$. It follows that $y - xc \in \Sigma L_\mu R$ and so $y \in \Sigma L_\mu R$. Next if y is impure, then since $y \notin L_o$, some monomial u in the support of y is the leading term of some pure element x of L. It follows that $h(x) < h(y)$, therefore $x \in \Sigma L_\mu R$, and again for some $c \in R_o$ the support of $y - xc$ does not contain u, so $h(y - xc) < h(y)$ and $y \in \Sigma L_\mu R$. So in either case we have reached a contradiction, and this proves that $L = \Sigma L_\mu R$.

For the proof that the map $\oplus L_\mu \otimes R \longrightarrow L$ is injective we shall isolate the following evident properties of the family $\{L_\mu\}$:

For each $\mu \in \Lambda \cup \{0\}$,

A_μ. *All elements of* L_μ *are* μ*-pure.*

For all $\mu_1, \mu_2 \in \Lambda \cup \{0\}$,

$B_{\mu_1\mu_2}$. *The* μ_1*-support of* L_{μ_1} *contains no monomial* u *which is also the* μ_1*-leading term of a (non* μ_1*-pure) element* xa, *where* $x \in L_{\mu_2}$, $a \in R$ *and if* $\mu_1 = \mu_2$, $\deg xa > \deg x$ *as well.*

Given $\mu \in \Lambda \cup \{0\}$, we choose for each monomial u that is the leading term of some element of L_μ an element $q \in L_\mu$ having this leading term with coefficient 1, and denote the set of all such q's by Q_μ. From the well-ordering of U and property A_μ it follows that Q_μ is an R_o-basis of L_μ.

Let $\lambda \in \Lambda$ and for each $u \in U^\lambda$ that is the λ-leading term of an element of L_o choose an element $q \in L_o$ having u as λ-leading term with coefficient 1, and denote the set of all such q's by $Q_{o\lambda}$. By property A_o every element of R_o has a λ-leading term, hence by the well-ordering $Q_{o\lambda}$ is an R_o-basis of L_o for each $\lambda \in \Lambda$. We shall call the elements of Q_λ "associated with λ", those of $Q_{o\lambda}$ "not associated with λ", and write $Q = \cup(Q_{o\lambda} \cup Q_\lambda)$. Further, V will denote the union of Q_o and the set of all products

$$qt_1 \ldots t_n, \quad q \in Q, \quad t_i \in T, \quad n \geq 1,$$

where no two successive terms are associated with the same index; thus if $q \in Q_{o\lambda}$ then $n \geq 1$, and $t_1 \in T_\lambda$.

We shall show that the elements of V have distinct leading terms and hence are right R_o-independent. By the same argument as in the proof of Th.5.1.2 we conclude that $\oplus L_\mu \otimes \cong VR_o = L$ and the proof will be complete.

From the lexicographic ordering of U it follows that if the choice of $q \in Q$ was determined by the monomial u, then $qt_1 \ldots t_n$ has leading term $ut_1 \ldots t_n$. Thus we are led to consider an equality of the form

$$ut_1 \ldots t_m = u't'_1 \ldots t'_n \qquad \text{in } U;$$

if $m \geq n$ say, this reduces to an equality of the form

$$ut_1 \ldots t_{m-n} = u'.$$

Let $q \in L_{\mu_2}$, $q' \in L_{\mu_1}$ correspond to u, u' respectively; there are two cases:

Case 1, $m > n$. If $\mu_1 \neq 0$, then $t_{m-n} \in T_{\mu_1}$, so the μ_1-support of $q' \in L_{\mu_1}$ contains $ut_1 \ldots t_{m-n-1}$ which is also the μ_1-leading term of the non μ_1-pure element $qt_1 \ldots t_{m-n-1}$, and this contradicts $B_{\mu_1\mu_2}$. If $\mu_1 = 0$, then since the support of $q' \in L_o$ contains $ut_1 \ldots t_{m-n}$ which is also the leading term of the pure element $qt_1 \ldots t_{m-n}$, we again have a contradiction to $B_{\mu_1\mu_2}$.

Case 2; $m = n$, then $u = u'$ is associated with λ, say. Then q and q' belong to $Q_{o\mu} \cup Q_\lambda$, where μ is the index associated with $t_1 = t'_1$ if $m > 0$, and $\mu = 0$, $Q_{o\mu} = Q_o$ if $m = 0$. By the construction of the Q's, if $q \neq q'$, they cannot belong to the same set, say $q \in Q_{o\mu}$, $q' \in Q_\lambda$. Then the support of $q \in L_o$ contains a monomial u which is also the leading term of a pure element $q'.1$, where $q' \in L_\lambda$, and this contradicts $B_{o\lambda}$.

This shows that the elements of V have distinct leading terms, hence V is an R_o-basis of L, as we wished to show. ■

Corollary 1 *(Bergman* [74]*). Let* $\{R_\lambda\}$ *be any family of non-trivial* R_o*-rings, where* R_o *is a skew field, and* $R = \underset{R_o}{\sqcup} R_\lambda$*, then*

$$\text{r.gl.dim.}R = \sup_\lambda\{\text{r.gl.dim.}R_\lambda\},$$

when the right-hand side is positive, and r.gl.dim.R $\leq$ 1 *when all the* R_λ *have* r.gl.dim. 0.

Proof. By Prop.5.1.3, r.gl.dim.R $\geq \sup_\lambda\{\text{r.gl.dim.}R_\lambda\}$. Now let M be a submodule of a free right R-module F, then by Th.5.2.1, $M \cong \oplus M_\mu \otimes R$, where each M_μ is an R_μ-submodule of F, which is projective as R_μ-module. Hence $\text{hd}_{R_\mu} M_\mu \leq$ (r.gl. dim.R_μ) - 1, or 0, if this is negative; so again by Prop. 5.1.3, $\text{hd}_R M \leq$ sup {r.gl.dim.R} - 1 or 0 if this is negative. ■

Corollary 2 *(Bergman* [74]*). With the notation of Cor.*1*, every projective* R*-module has the form* $\oplus P_\mu \otimes_{R_\mu} R$*, where each* P_μ *is projective as* R_μ*-module.*

Proof. If P is a projective R-module, then P is a submodule of a free R-module, so by Th.5.2.1, $P \cong \oplus P_\mu \otimes R$ and each P_μ is projective as right R_μ-module, by Prop.5.1.3. ■

5.3 The monoid of projectives

Let R be any ring and write $\mathcal{P}(R)$ for the set of isomorphism types of finitely generated projective right R-modules. The element of $\mathcal{P}(R)$ corresponding to the module P will be written [P], so that [P] = [P'] if and only if $P \cong P'$, by definition. On $\mathcal{P}(R)$ we define a binary operation by the rule

$$[P] + [Q] = [P \oplus Q].$$

Since $P \cong P'$, $Q \cong Q'$ $\Rightarrow P \oplus Q \cong P' \oplus Q'$, this operation is well-defined and it is clear that $P(R)$ becomes a monoid in this way, with $[0]$ as neutral element. Moreover, any ring-homomorphism $f: R \longrightarrow S$ induces a monoid homomorphism $P(R) \longrightarrow P(S)$, defined by $[P] \longmapsto [P \otimes_R S]$, where S is a left R-module by pullback along f. In this way P becomes a functor from rings to monoids. Our objective will be to show that P preserves coproducts over skew fields. More precisely we have

Theorem 5.3.1 *(Bergman [74]). Let* R_o *be a skew field,* $\{R_\lambda\}$ *a family of non-trivial* R_o*-rings and* R *their coproduct, then the map*

$$(1) \quad \coprod_{P(R_o)} P(R_\lambda) \longrightarrow P(R)$$

induced in the category of monoids is an isomorphism.

Proof. We first prove injectivity. Thus assume that two distinct elements $\Sigma[L_\mu]$, $\Sigma[M_\mu]$ of the left-hand side of (1) have the same image, i.e. there is an isomorphism

$$(2) \quad \alpha: \oplus L_\mu \otimes R \longrightarrow \oplus M_\mu \otimes R.$$

We shall retain the notation of the proofs of Th.5.1.2 and 5.2.1 for $M = \oplus M_\mu \otimes R$ and identify L_μ with its image $L_\mu\alpha$ in M. Since each L_μ is finitely generated, we can associate with the map α an element $h_\alpha \in H$ defined by the rule

$$h_\alpha(\mu,u) = \begin{cases} 1 \text{ if } u \text{ is in the } \mu\text{-support of } L_\mu \text{ (and so } u \in U^\mu), \\ 0 \text{ otherwise.} \end{cases}$$

We may assume the pair $\Sigma[L_\mu] \neq \Sigma[M_\mu]$ and the isomorphism α in (2) to be chosen so as to minimize h_α. With this choice we claim that the family $\{L_\mu\}$ satisfies the conditions A_μ,

$B_{\mu_1\mu_2}$ of the proof of Th.5.2.1, and hence $L_\mu \alpha = M_\mu$. It will be important for later applications that this can be done without using the finite generation of the M_μ nor the projectivity of the L_μ, M_μ.

If A_{μ_1} fails then some $x \varepsilon L_{\mu_1}$ is not μ_1-pure. Let $u \varepsilon U^{\mu_1}$ be the μ_1-leading term of x and consider the restriction $p'_{\mu_1 u} = p_{\mu_1 u}|L_{\mu_1}$ (where $p_{\mu_1 u}:M \longrightarrow R_{\mu_1}$ is as defined on p.99). The image of x is a non-zero element of R_o, hence $p'_{\mu_1 u}$ is surjective and so splits; thus $L_{\mu_1} = L'_{\mu_1} \oplus xR_{\mu_1}$, where $L'_{\mu_1} = \ker p'_{\mu_1 u}$ and u is clearly not in the μ_1-support of L'_{μ_1}. Now x must be μ_2-pure for some $\mu_2 \neq \mu_1$. Taking a new symbol $\bar{x}$, we define $L'_{\mu_2} = L_{\mu_2} \oplus \bar{x}R_{\mu_2}$ and for $\mu \neq \mu_1,\mu_2$ put $L'_\mu = L_\mu$. There is an obvious isomorphism $\oplus L'_\mu \otimes R \longrightarrow \oplus L_\mu \otimes R$ sending $\bar{x}$ to x, and if we follow this by α we obtain an isomorphism β: $\oplus L'_\mu \otimes R \longrightarrow M$. The first place where h_β differs from h_α is either (μ_1,u) or (μ_2,u'), where $u \varepsilon u'R_{\mu_2}$. If $\mu_2 \neq 0$, then $(\mu_1,u) > (\mu_2,u')$ by considering degrees, and for $\mu_2 = 0$ this inequality still holds, by the ordering by first components. Thus $h_\alpha(\mu_1,u) = 1 > 0 = h_\beta(\mu_1,u)$, which contradicts the minimality of h_α, because $\Sigma[L'_\mu] = \Sigma[L_\mu] \neq \Sigma[M_\mu]$. Hence A_μ holds for all $\mu \varepsilon \Lambda \cup \{0\}$.

If $B_{\mu_1\mu_2}$ fails, let the μ_2-support of L_{μ_1} contain a monomial u which is also the μ_1-leading term, with coefficient 1, of a (non μ_1-pure) element xa, where $x \varepsilon L_{\mu_2}$, $a \varepsilon R$ and if $\mu_1 = \mu_2$, deg xa > deg x. For each $y \varepsilon L_{\mu_1}$ there is a unique element $y\phi$ of M of the form $y\phi = y - xa(yp)$ $(yp \varepsilon R_{\mu_1})$, whose μ_1-support does not contain u. Indeed, $p = p'_{\mu_1 u}$: $L_{\mu_1} \rightarrow R_{\mu_1}$; we observe that this is R_{μ_1}-linear. Let $\phi:M \longrightarrow M$ be the R-linear map which leaves every element of L_μ $(\mu \neq \mu_1)$

fixed and which maps $y \in L_{\mu_1}$ to $y\phi$. We claim that ϕ is an automorphism, with inverse Θ leaving L_μ $(\mu \neq \mu_1)$ fixed and sending $y \in L_{\mu_1}$ to $y + xa(yp)$. If $\mu_1 \neq \mu_2$, then $x \in L_{\mu_2}$ is fixed by ϕ and Θ. If $\mu_1 = \mu_2$, then deg xa > deg x, so u is not in the μ_1-support of x and x is again left fixed by ϕ and Θ. It follows that ϕ and Θ are mutually inverse. Now consider the isomorphism $\gamma: \oplus L_\mu \otimes R \xrightarrow{\alpha} M \xrightarrow{\phi} M$. The first place where h_γ and h_α differ is (μ_1, u) and it is clear that $h_\gamma < h_\alpha$. This contradiction shows that $B_{\mu_1\mu_2}$ holds. Now we can construct V as in the proof of Th.5.2.1.

We now complete the proof by showing that $L_\mu = M_\mu$ for all μ. If $x \in M_{\mu_1}$, then x has degree 1; write $x = \Sigma v x_v$ $(x_v \in R)$ and choose $v \in V$ such that $x_v \neq 0$. Then the leading term of v must have degree 1, so v has length 1 as element of V, so v lies in some L_{μ_2}. Hence v is μ_2-pure, but is of degree 1, so $v \in M_{\mu_2}$. Since the elements of V are R_o-independent, we conclude that $\mu_1 = \mu_2$, i.e. $M_{\mu_1} \subseteq L_{\mu_1}$ (i.e. $L_{\mu_1}\alpha$, to be precise). But R is free as left R_{μ_2}-module, so equality must ho
Thus $M_\mu = L_\mu$ for all μ; this proves (1) to be injective. To establish surjectivity, let P be a finitely generated project R-module, say $P \oplus Q \cong R^n$. By Th.5.2.1, Cor.2, $P \cong \oplus P_\mu \otimes R$, $Q \cong \oplus Q_\mu \otimes R$, and it remains to show that the P_μ are finitely generated. There is an isomorphism $\alpha: \oplus R_\mu^{n_\mu} \otimes R \longrightarrow \oplus (P_\mu \oplus$ such that $\Sigma n_\mu = n$; if we choose the n_μ and α so as to minimis then as in the first part of the proof $R_\mu^{n_\mu} \cong P_\mu \oplus Q_\mu$, hence P itely generated. Thus (1) is surjective, and hence an isomorp

This result enables us to derive several useful consequences without difficulty (cf. Cohn [63,64,68], Bergman [74]

Theorem 5.3.2. *The coproduct of a family of firs over a skew field is a fir. In particular, the coproduct of skew fields (over a skew field) is a fir.*

Proof. By Th.5.2.1, Cor.1, the ideals of the coproduct are

projective, and by Th.5.3.1 all projectives are free of unique rank, hence all ideals are free of unique rank. ■

In 4.3 we saw that any semifir (and in particular, any fir) has a universal field of fractions in which all full matrices become invertible. So we conclude that the coproduct of fields E_λ over a field K has a universal field of fractions. This will be called the *field coproduct* or simply *coproduct* of the E_λ, written $\overset{o}{_K} E_\lambda$ or in the case of two factors E,F, $E \overset{o}{_K} F$. Let us recall from Cohn [71"], ch.1 that a ring is Morita-equivalent to a fir if and only if it is an n x n matrix ring over a fir, and observe that Th.5.2.1. Cor.1,2 and Th.5.3.2 are statements about categories of modules and hence Morita-invariant. We deduce

Theorem 5.3.3 *(Bergman* [74]*). Let* R_o *be an* n x n *matrix ring over a skew field,* $\{R_\lambda\}$ *a family of faithful* R_o*-rings and* R *their coproduct over* R_o*. Then*

(i) $\text{r.gl.dim.}R = \begin{cases} \sup_\lambda\{\text{r.gl.dim.}R_\lambda\} & \textit{if this is} \neq 0, \\ 0 \textit{ or } 1 \textit{ otherwise;} \end{cases}$

(ii) *every projective* R*-module has the form* $\oplus P_\mu \otimes R$*, where* P_μ *is* R_μ*-projective,*

(iii) $P(R) = \coprod_{P(R_o)} P(R_\lambda)$. ■

If we examine the proof of Th.5.3.1, we find that it shows rather more than is asserted. Consider a homomorphism

(3) $\quad \phi: \oplus M_\mu \otimes R \longrightarrow \oplus N_\mu \otimes R.$

If ϕ arises from a family of R_μ-linear maps $\phi_\mu : M_\mu \longrightarrow N_\mu$, we shall call it *induced*. Next we observe that

(4) $\quad (M_{\mu_1} \oplus R_{\mu_1}) \otimes R \oplus M_{\mu_2} \otimes R \cong M_{\mu_1} \otimes R \oplus (M_{\mu_2} \oplus R_{\mu_2}) \otimes R,$

because $R_{\mu_1} \otimes R \cong R_{\mu_2} \otimes R \cong R$. An isomorphism (3) arising by a transfer of terms as in (4) is called a *free transfer isomorphism*. A second kind of isomorphism arises as follows. Let $e:M_{\mu_1} \longrightarrow R_{\mu_1}$ be an R_{μ_1}-linear functional, extended to M so as to annihilate M_μ for $\mu \neq \mu_1$. Given $x \in M_{\mu_2}$, we have a map $\alpha(x):R \longrightarrow M$ defined by $1 \longmapsto x \in M_{\mu_2} \otimes R \subseteq M$. Clearly $\alpha(x)e = 0$ if $\mu_1 \neq \mu_2$ and this holds even for $\mu_1 = \mu_2$ if we add the condition $x \in \ker e$. Then for any $a \in R$ the map $e\alpha(a)\alpha(x):u \longmapsto xa(ue)$ is nilpotent, and so $\Theta = 1 - e\alpha(a)\alpha(x)$ is an automorphism of M; such an automorphism will be called a *transvection*, μ-based in case $\mu_1 = \mu_2 = \mu$ and $a \in R_\mu$. The proof of Th.5.3.2 shows that every surjection (3) where M_μ is finitely generated can be obtained as a composite of a finite number of free transfer isomorphisms, transvections and an induced surjection.

Proposition 5.3.4. *Let* R_o *be a skew field and* $\{R_\lambda\}$ *a family of entire* R_o*-rings, then their coproduct* R *is entire and any unit in* R *is a product of units in the* R_λ.

We shall call such a unit a *monomial unit*.

Proof. Each unit $u \in R$ defines an automorphism $x \longmapsto ux$ of $R = R_o \otimes R$. Here R is free on a single generator; any free transfer isomorphism just amounts to renaming the generator, while a surjection is a unit in some R_μ. The only transvection is the identity map, since it must be μ-based for some μ, but R_μ is entire. This proves the second part. Now the assertion about zero-divisors is the special case n = 1 of the next result.

Proposition 5.3.5. *Let* R_o *be a skew field and* n *a positive integer. Then the coproduct of any family of* n*-firs over* R_o *is an* n*-fir*.

Proof. For any map $R^n \longrightarrow R$ the image can by Th.5.2.1 be written as $\oplus M_\mu \otimes R$, so there is an induced surjection $\alpha: \oplus R_\mu^{n_\mu} \otimes R \longrightarrow \oplus M_\mu \otimes R$, where $\Sigma n_\mu = n$. Since M_μ is a submodule of the projective R_μ-module R, M_μ is free of rank at most n_μ and so $\oplus M_\mu \otimes R$ is free of rank at most n. If we repeat this argument with an isomorphism, we see that the rank must be unique. ■

Bergman [74] determines the units and zero-divisors of coproducts over skew fields in a more general situation.

We can also obtain the elements of a coproduct which are algebraic over the ground field. Let $\{R_\lambda\}$ be a family of entire rings over a skew field R_o; their coproduct R is a free product and is entire, by Th.5.1.2, Cor. and Prop.5.3.4. If $a \in R$ is right algebraic over R_o, it satisfies an equation

$$a^n\gamma_o + a^{n-1}\gamma_1 + \dots + \gamma_n = 0, \qquad \gamma_i \in R_o, \text{ not all } 0.$$

Without loss of generality $a \neq 0$, and since R is entire, we may assume that $\gamma_n \neq 0$; on dividing by the constant term we may take $\gamma_n = 1$. Then $a(a^{n-1}\gamma_o + \dots + \gamma_{n-1}) + 1 = 0$, i.e. $ab + 1 = 0$, so $a(-b) = 1$ and it follows that a is a unit in R. By Prop.5.3.4, it must be a monomial unit and we can write it as $a = p^{-1}up$, where either $u \in R_\lambda$ for some λ, or if $\deg u > 1$, the first and last monomial factors of u are in different rings R_λ. In the latter case we may also assume that $\deg up = \deg u + \deg p$; then

$$(5) \qquad u^np\gamma_o + u^{n-1}p\gamma_1 + \dots + up\gamma_{n-1} + p = 0.$$

But $\deg(u^rp) = r.\deg u + \deg p$, so the first term in (5) has greater degree than the rest and we have a contradiction. So this case can be excluded, and we obtain

Corollary 1. *Let* $\{R_\lambda\}$ *be a family of entire rings over a skew field* R_o *and* R *their coproduct. Then any element of* R *which is right (or left) algebraic over* R_o *is conjugate to an element in one of the factors.* ■

It should be observed that not every element conjugate

to an element of some R_λ that is algebraic over R_o need itself be algebraic over R_o. For if a satisfies an equation over R_o, $p^{-1}ap$ may not do so; only when R_o is in the centre of each R_λ and each R_λ algebraic over R_o does the converse of Cor.1 hold.

When different subfields are amalgamated in different factors Th.5.3.2 and Prop.5.3.4 no longer hold, in fact R need not even be entire. This follows from an example for groups due to B.H. Neumann [54]. Let k be a commutative field and form the fields

$$K_1 = k(x,y) \text{ with defining relation } y^{-1}xy = x^{-1},$$
$$K_2 = k(y,z) \text{ with defining relation } z^{-1}yz = y^{-1},$$
$$K_3 = k(z,x) \text{ with defining relation } x^{-1}zx = z^{-1}.$$

These fields can of course be constructed as fields of fractions of skew polynomial rings, e.g. $K_1 = L(y;\alpha)$, where $L = k(x)$ with automorphism $\alpha: f(x) \longmapsto f(x^{-1})$.

We form the coproduct P of K_1, K_2, K_3 with amalgamations $K_{12} = k(y)$, $K_{23} = k(z)$, $K_{31} = k(x)$. To see that this is a free product, we first form $k(\xi,\eta,\zeta)$ in three commuting indeterminates. Next adjoin x subject to

$$f(\xi,\eta,\zeta)x = xf(\xi,\eta,\zeta^{-1}), \qquad x^2 = \xi.$$

Then adjoin y subject to

$$f(x,\eta,\zeta)y = yf(x^{-1},\eta,\zeta), \qquad y^2 = \eta,$$

and finally adjoin z subject to

$$f(x,y,\zeta)z = zf(x\zeta,y^{-1},\zeta) \qquad z^2 = \zeta.$$

It is easily verified that the resulting coproduct is a

free product. If e.g. there is a relation between x and y, we can write it as a polynomial in y:

$$y^n + a_1 y^{n-1} + \dots + a_n = 0 \qquad a_i = a_i(x).$$

Conjugating by $\zeta = z^2$ we find

$$y^n + a_1(x\,\zeta^2)y^{n-1} + \dots + a_n(x\,\zeta^2) = 0,$$

and by the uniqueness of the minimal equation for y, $a_i(x\,\zeta^2) = a_i(x)$, hence a_i is independent of x, so y is algebraic over k, clearly a contradiction. Thus P is a free product, but in P, xyz is an element of order 2:

$$xyz = yx^{-1}z = yz^{-1}x^{-1} = z^{-1}y^{-1}x^{-1},$$

thus $(xyz)^2 = 1$; this shows that P is not even entire.

5.4 The tensor K-ring on a bimodule

Let K be any field, then the free K-ring on a set X, K<X>, may be defined by the following universal mapping property: K<X> is generated by X as a K-ring, and any mapping $X \longrightarrow R$ into a K-ring R such that the image of X centralizes K (i.e. ya = ay for all $a \in K$ and y in the image of X) can be extended to a unique K-ring homomorphism of K<X> into R. The elements of K<X> can be uniquely written as

$$\Sigma a_{i_1 \dots i_r} x_{i_1} \dots x_{i_r} \qquad (x_i \in X,\ a_{i_1 \dots i_r} \in K).$$

As is easily seen, K<X> may also be represented as a coproduct

$$K\langle X\rangle \cong \coprod_X K[x]\ ,$$

where x runs over X, and since each $K[x]$ is a principal ideal domain (and hence a fir), it follows from Th.5.3.2 that $K\langle X\rangle$ is a fir. We shall now outline another way of establishing this fact which will be useful when we come to consider a generalization of $K\langle X\rangle$ needed later.

Let K be any field and M a K-bimodule. We put

$$M^n = M \otimes M \otimes \ldots \otimes M \qquad \text{(n factors)},$$

thus $M^1 = M$ and by convention, $M^o = K$. It is clear that $M^r \otimes M^s \cong M^{r+s}$, hence we have a multiplication on the direct sum

$$\mathbf{T}(M) = \oplus M^n,$$

which turns it into a K-ring. This ring is called the *tensor K-ring* on M. It is shown in Cohn [71"], Ch.2 that this ring possesses a weak algorithm and hence is a fir. We shall not repeat the proof here, but note that it yields another proof that $K\langle X\rangle$ is a fir. Let $M = {}^X K = \underset{X}{\oplus} K$ and denote by x the element corresponding to $1 \in K$ in the factor indexed by $x \in X$. Thus the general element of M has the form $\Sigma a_x x$ ($a_x \in K$, almost all are 0), and it is easily seen that $\mathbf{T}(M) = K\langle X\rangle$.

Let E be a field with K as subfield, and put

$$(1) \qquad E_K\langle X\rangle = E \underset{K}{\sqcup} K\langle X\rangle.$$

By Th.5.3.2 this is again a fir, but we can also obtain this ring as the tensor E-ring of a module. In $E_K\langle X\rangle$ consider the E-subbimodule spanned by $x \in X$. Its elements are of the form $\Sigma a_i x b_i$ ($a_i, b_i \in E$) and it is clear from the definition that this module is isomorphic to $E \otimes_K E$. Hence we see that $E_K\langle X\rangle$ can be described as a tensor E-ring as follows:

$$E_K\langle X\rangle = \mathbf{T}({}^X(E \otimes_K E)).$$

Since this is a fir (hence a semifir) we see by Th.4.C that it has a universal field of fractions. We shall write

$$E_K\{X\} = \mathbf{F}(E_K\langle X\rangle).$$

When E = K, the ring $K_K\langle X\rangle$ is just the tensor ring $K\langle X\rangle$ introduced earlier, so no confusion need arise if we regard $K\langle X\rangle$ as an abbreviation for $K_K\langle X\rangle$, and correspondingly denote the universal field of fractions by $K\{X\}$.

To elucidate the relation between $E_K\{X\}$ and $K\{X\}$ we need a lemma.

Lemma 5.4.1. *Let* R, S *be semifirs over a field* K, *then* (i) *the inclusion map* $R \longrightarrow R \sqcup S$ *is honest,* (ii) *the map* $R \sqcup S \longrightarrow \mathbf{F}(R) \sqcup S$ *is honest.*

Proof. (i) Consider the injections

$$R \longrightarrow R \sqcup S \longrightarrow \mathbf{F}(R) \sqcup S.$$

Let A be full over R, then it is invertible over $\mathbf{F}(R)$, hence also over $\mathbf{F}(R) \sqcup S$, and so is full over $R \sqcup S$, as we had to show.

(ii) By (i) any full matrix over R is full over $R \sqcup S$, hence we have a homomorphism $\mathbf{F}(R) \longrightarrow \mathbf{F}(R \sqcup S)$ (Th.4.3.3) and it follows that we have a homomorphism $\mathbf{F}(R) \sqcup S \longrightarrow \mathbf{F}(R \sqcup S)$. Thus we have mappings

$$R \sqcup S \longrightarrow \mathbf{F}(R) \sqcup S \longrightarrow \mathbf{F}(R \sqcup S).$$

Now any full matrix over $R \sqcup S$ is invertible over $\mathbf{F}(R \sqcup S)$ and hence is full over $\mathbf{F}(R) \sqcup S$. ■

Proposition 5.4.2. *Let* $K \subseteq E$ *be any fields, then*

$$E\{X\} \cong E \circ_K K\{X\}.$$

Proof. Put $R = K\langle X\rangle$, then we have to show that

(2) $\quad \mathbf{F}(R \sqcup E) = \mathbf{F}(R) \underset{K}{\circ} E.$

By the lemma, the map $R \sqcup E \longrightarrow \mathbf{F}(R) \sqcup E$ is honest, so $\mathbf{F}(R \sqcup E)$ is embedded in $\mathbf{F}(\mathbf{F}(R) \sqcup E)$, but $\mathbf{F}(R \sqcup E)$ is a field containing both R and E, hence we have equality. ■

Corollary. Let $K \subseteq E \subseteq F$ *be any fields, then*

$$E_K\langle X\rangle \subseteq F_K\langle X\rangle.$$

Proof. Write $L = K\langle X\rangle$, then we have the sequence of maps

$$E \sqcup L \longrightarrow F \sqcup L \longrightarrow F \sqcup \mathbf{F}(E \sqcup L).$$

Any full matrix over $E \sqcup L$ is invertible over $F \sqcup \mathbf{F}(E \sqcup L)$, and hence full over $F \sqcup L$. Thus $\mathbf{F}(E \sqcup L)$ is embedded in $\mathbf{F}(F \sqcup L)$ and by Prop.5.4.2, this is the result claimed. ■

Suppose that we have fields $K \subseteq E \subseteq F$, $K \subseteq K' \subseteq F$, then we have a natural map

(3) $\quad E_K\langle X\rangle \longrightarrow F_{K'}\langle X\rangle,$

and when this map is honest, we obtain an embedding of $E_K\langle X\rangle$ in $F_{K'}\langle X\rangle$, but in general (3) need not even be injective. Thus let $x \in X$ and $c \in K' \cap E$, then $cx - xc$ maps to 0 under (3), but it is not itself 0 unless $c \in K$. Thus a necessary condition for (3) to be injective is that

(4) $\quad K' \cap E = K.$

Later in 6.3 we shall see that when K is contained in the centre of E, (4) is also sufficient for (3) to be honest.

5.5 Subfields of field coproducts

Although Schreier had discussed free products of groups

in 1927, it was not until more than 20 years later that significant applications were made, notably in the classic paper by Higman-Neumann-Neumann [49]. Their main result was the following

Theorem 5.A. *Let* G *be any group with two subgroups* A, B *which are isomorphic, say* $f: A \longrightarrow B$ *is an isomorphism. Then* G *can be embedded in a group* H *containing also an element* t *such that*

$$t^{-1}at = af \qquad \textit{for all } a \in A.$$

We observe that this would be trivial if f were an automorphism of the whole of G: then H would be the split extension of G by an infinite cycle inducing f. But for proper subgroups A, B the result is non-trivial and (at first) surprising. It has many interesting and important consequences for groups and it is natural to try and prove an analogue for skew fields. What one needs is a coproduct in the category of fields. However, we shall not adopt a categorical point of view: the morphisms in the category of fields are all monomorphisms and this is somewhat restrictive. Over a fixed ring, it is true, we have defined specializations, but it would be more cumbersome to define them without a ground ring, and not really helpful.

In this section we shall prove an analogue of the Higman-Neumann-Neumann theorem using the field coproduct introduced in 5.3. But we shall also need some auxiliary results on subfields of coproducts. It will be convenient to regard all our fields as algebras over a given commutative field k; this just amounts to requiring k to be contained in the centre of each field occurring. The proof of the next result is based on a suggestion by A. Macintyre.

Theorem 5.5.1. *Let* K *be a field and* A, B *subfields of* K, *isomorphic under a mapping* $f: A \longrightarrow B$, *where* K,A,B *are* k-

algebras and f *is* k*-linear. Then* K *can be embedded in a field* L*, again a* k*-algebra, in which* A *and* B *are conjugate by an inner automorphism inducing* f*, i.e.* L *contains* $t \neq 0$ *such that*

$$af = t^{-1}at \qquad \textit{for all } a \in A.$$

Proof. Define K as right A-module by the usual multiplication and as left A-module by

(1) $a.u = (af)u \qquad a \in A,\ u \in K.$

Let us form the K-bimodule $K \otimes_A K$, with the usual multiplication by elements of K; if we abbreviate $1 \otimes 1$ as t, this consists of all sums $\Sigma u_i t v_i$ $(u_i, v_i \in K)$ with the defining relations

(2) $at = t.af \qquad (a \in A).$

By the remarks in 5.4, the tensor K-ring $T(K \otimes_A K)$ is a fir, and so has a universal field of fractions L. Thus we have embedded K in a field L in which (2) holds. ∎

Let K be a field with k as subfield of the centre, then K is said to be *finitely homogeneous* over k, if for any elements $a_1,\ldots,a_n,b_1,\ldots,b_n \in K$ such that the map $a_i \mapsto b_i$ defines an isomorphism $k(a_1,\ldots,a_n) \cong k(b_1,\ldots,b_n)$, there exists $t \in K^*$ such that $t^{-1}a_i t = b_i$ $(i = 1,\ldots,n)$.

Corollary 1. *Every field* K *(over a subfield* k *of its centre) can be embedded in a field (again over* k*) which is finitely homogeneous.*

Proof. Given a's and b's such that $a_i \mapsto b_i$ defines an isomorphism, we can by Th.5.5.1 extend K to include an element $t \neq 0$ such that $t^{-1}a_i t = b_i$, and the least such extension has the same cardinal as K or is countable. If we do this for all pairs of finite sets in K which define isomorphisms we get a field K_1, still of the same cardinal as K or countable, such that any two finitely generated isomorphic subfields of K are conjugate in K_1. We now repeat this process, obtaining K_2 and if we continue thus we get a tower of fields

$$K \subset K_1 \subset K_2 \subset \dots$$

Their union L is a field with the required properties, for if $a_1,\dots,a_n$, $b_1,\dots,b_n \in L$ and $a_i \longmapsto b_i$ defines an isomorphism, we can find K_r to contain all the a's and b's, hence they become conjugate in K_{r+1} and a fortiori in L.■

A finitely homogeneous field has the property that any two elements with the same (or no) minimal equation over k are conjugate. Hence we obtain

Corollary 2. *Every field* K *(over a subfield* k *of its centre) can be embedded in a field* L *in which any two elements with the same minimal equation over* k *are conjugate, as are any two transcendental elements.*■

Let K be any field, then the group of fractional linear transformations $PGL_2(K)$, consisting of all mappings

$$x \longmapsto (ax + b)(cx + d)^{-1} \qquad \left(\begin{pmatrix} a & b \\ c & d \end{pmatrix} \in GL_2(K) \right)$$

is well known to be triply transitive, i.e. for any two triples of distinct elements of $K_\infty = K \cup \{ \infty \}$ there is a transformation mapping one into the other. With the help of Cor.1 we can (as P.J. Cameron has observed) construct a field on which PGL_2 is 4-transitive. We need only take the field of rational functions in one variable over GF(2) and embed it in a finitely homogeneous field L. Given any two elements a,b of L different from 0 and 1, there is an inner automorphism σ mapping a to b. Now by the result quoted above, PGL_2 is triply transitive on L_∞ and the stabilizer of $\{0,1,\infty\}$ is still 4-transitive (by conjugation), hence PGL_2 is 4-transitive.

Later we shall need an analogue of Cor.2 for matrices instead of elements. Let $A \in \mathfrak{M}_n(K)$, then A is said to be *transcendental* over k if for every non-zero polynomial $f \in k[t]$ the matrix f(A) is non-singular. Clearly if A is a transcendental matrix over k, then the field generated over k by A is a simple transcendental extension of k. The next lemma and its application to the proof of Th.5.5.3 are due to G.M. Bergman (cf. Cohn [73"]).

Lemma 5.5.2. *Given a field* K *(over* k*) and* $n \geq 1$, *let* E *be a subfield of* $\mathfrak{M}_n(K)$. *If* F_1, F_2 *are subfields of* E *which are isomorphic under a map* $\phi: F_1 \longrightarrow F_2$, *then there is an extension field* L *of* K *such that*

$$x\phi = T^{-1}xT \qquad \textit{for all } x \in F_1,$$

for some $T \in GL_n(L)$.

Proof. By Th.5.5.1 E has an extension field E' with an element T inducing ϕ. Consider $R = \mathfrak{M}_n(K) \underset{E}{\sqcup} E'$; by Th.5.3.3 this is hereditary and every projective R-module is a direct sum of copies of $P \otimes R$, where P is a minimal projective for $\mathfrak{M}_n(K)$. Since $P^n \otimes R = \mathfrak{M}_n(K) \otimes R \cong R$, it follows (by Th.1.4. of Cohn [71"]) that R is an n x n matrix ring over a fir, say $R = \mathfrak{M}_n(G)$, where G is a fir containing K. Let L be the universal field of fractions of G, then L contains K and $\mathfrak{M}_n(L)$ contains the element T inducing ϕ. ■

Let K be a field and suppose that $\mathfrak{M}_n(K)$ contains isomorphic subfields F_1,F_2,F_3 with isomorphisms $f:F_1 \longrightarrow F_2$, $g:F_2 \longrightarrow F_3$ say, and such that F_1, F_2 lie in a common subfield of $\mathfrak{M}_n(K)$, as do F_2, F_3. Then by Lemma 5.5.2 we can enlarge K to a field L and obtain a unit X such that conjugation by X induces f. Now F_2, F_3 still lie in a common subfield of $\mathfrak{M}_n(L)$ and we can enlarge L to a field M to obtain a unit Y which induces the isomorphism g between F_2 and F_3. Now XY induces the isomorphism $fg : F_1 \longrightarrow F_3$; in this way the scope of the lemma can be extended. As a result we can prove

Theorem 5.5.3. *Let* K *be a field (over* k*) and* $n \geq 1$. *Given two* $n \times n$ *matrices* A, B *over* K, *both transcendental over* k, *there exists a field extension* L *of* K *containing a non-singular matrix* T *such that* $T^{-1}AT = B$.

Proof. Since A is transcendental, k(A) is a purely transcendental extension of k, thus if u is a central indeterminate over K, we have $k(A) \cong k(u)$ and similarly $k(B) \cong k(u)$. We shall take $F_1 = k(A)$, $F_2 = k(u)$, $F_3 = k(B)$. Let K((u)) be the field of formal Laurent series in u over K, then

$$(3) \quad \mathfrak{M}_n(K((u))) \cong \mathfrak{M}_n(K)((u)).$$

Now F_1, F_2 are contained in the subfield k(A)((u)) of (3), while F_2, F_3 are contained in k(B)((u)). We can therefore apply Lemma 5.5.2 and the remark following it and obtain an extension field H of K((u)) such that $\mathfrak{M}_n(H)$ contains a unit T inducing the k-isomorphism $k(A) \cong k(B)$ defined by $A \longmapsto B$. ■

Clearly we can repeat the process for other pairs of transcendental matrices until we obtain a field $K_1 \supset K$ in which any two transcendental matrices of the same order over K are similar. If we repeat the construction for K_1 we get a chain of fields (over k):

$$K \subset K_1 \subset K_2 \subset \ldots$$

whose union is a field with the property that any two transcendental matrices of the same order are similar, thus we have the

Corollary. Let K *be a field (over* k*) then there exists an extension field* L *of* K *(over* k*) such that any two matrices of the same order over* L *and both transcendental over* k *are similar over* L. ■

This means, for example, that over L any transcendental matrix A can be transformed to *scalar* (not merely diagonal) form. We need only choose a transcendental element α of L; clearly α is transcendental as n x n matrix, therefore $T^{-1}AT = \alpha$ for some $T \in GL_n(L)$. We shall return to this topi in 8.4.

Our next objective is to show that every countable field can be embedded in a 2-generator field. This corresponds to a theorem of B.H. Neumann [54] for groups. We shall need some lemmas on field coproducts. First we examine a situation in which a subfield of a given field is a field coproduct.

Lemma 5.5.4. *Let* K *be a field with a subfield* H *and let* $P = K \circ_H H(x)$*, where* x *is an indeterminate centralizing* H. *Then the subfield* G *of* P *generated by the fields* $K_i = x^{-i}Kx^i$ $(i \in \mathbf{Z})$ *is their field coproduct over* H.

Proof. Take a family of copies of K indexed by **Z**, say $\{K_i\}$, denote by R their coproduct over H and by U the universal field of fractions of R, thus U is the field coproduct of the K_i over H. By the universal property of U it follows that the subfield Q described in the lemma is an R-specialization of U. From the universal property of $P = K \circ_H H(x)$, this specialization will be an isomorphism whenever there is some K-field L containing an element $y \neq 0$ such that the specialization from U to L which maps K_i to y^iKy^{-i} is an embedding. Such an L is easily constructed: the mapping $K_i \longrightarrow K_{i+1}$ is an automorphism of R which extends to an automorphism α say, of U. Now form the skew function field $U(y;\alpha)$; it has all the properties required of L. ■

We shall also need a result on free sets in field coproducts. Given a field over k, by a *free set* over k we understand a subset Y such that the subfield generated by Y is free, i.e. isomorphic to the universal field of fractions of the free algebra k<Y>.

Lemma 5.5.5. *Let* E *be a field, generated over* k *by a family* $\{e_\lambda\}$ *of elements, and let* U *be the field freely generated by a family* $\{u_\lambda\}$ *over* k, *then the elements* $u_\lambda + e_\lambda$ *form a free set in the field coproduct* $U \overset{o}{\underset{k}{}} E$.

Proof. The field coproduct $G = U \overset{o}{\underset{k}{}} E$ has the following universal property: given any E-field F and any family $\{f_\lambda\}$ of elements of F, there is a unique specialization from G to F (over E, with domain generated by E and the u_λ) which maps u_λ to f_λ. In particular, there are specializations from G to itself which map u_λ to $u_\lambda + e_\lambda$ (respectively to $u_\lambda - e_\lambda$). On composing these mappings (in either order) we obtain the identity mapping, hence they are inverse to each other, and so are automorphisms. It follows that the $u_\lambda + e_\lambda$ like the u_λ form a free set. ∎

We can now achieve our objective, the embedding theorem mentioned earlier; the proof runs closely parallel to the group case.

Theorem 5.5.6. *Let* E *be a field, countably generated over a subfield* k *of its centre, then* E *can be embedded in a 2-generator field over* k.

In essence the proof runs as follows: Suppose that E is generated by $e_o = 0, e_1, e_2, \ldots$; we construct an extension field L generated by elements x,y,z over E satisfying

$$y^{-i}xy^i = z^{-i}xz^i + e_i \qquad (i = 0,1,\ldots).$$

Then L is in fact generated by x,y,z alone. If we now adjoin t such that $y = txt^{-1}$, $z = t^{-1}xt$, the resulting field is generated by x and t.

To prove the theorem, let F_1 be the free field on x,y over k; it has a subfield U generated by $u_i = y^{-i}xy^i$ ($i = 0,1,\ldots$) freely, by Lemma 5.5.4, and similarly, let F_2 be the free field on x,z over k, with subfield V freely generated by $v_i = z^{-i}xz^i$ ($i = 0,1,\ldots$). Form $K = E \circ_k F_1$; this has a subfield W generated by $w_i = u_i + e_i$ ($i = 0,1,\ldots$), freely by Lemma 5.5.5. We note that $w_o = u_o + e_o = x_o = x$, so K is generated over k by x,y and the w_i ($i \geq 1$).

Let L be the field coproduct of K and F_2, amalgamating W and V along the isomorphism $w_i \longleftrightarrow v_i$. We note that $w_o = x = v_o$ and that L is generated by x,y,z and the w_i or also by x,y,z and the v_i, or simply by x,y,z. Now L contains the isomorphic subfields generated by x,y and by z,x respectivel hence we can adjoin t to L such that $t^{-1}xt = z$, $t^{-1}yt = x$ (by Th.5.5.1). It follows that we have an extension of L generated by x,t over k and it contains K. ■

As usual we have the

Corollary. Every field over k *can be embedded in a field* L *such that every countably generated subfield of* L *is contain in a 2-generator subfield of* L.

Proof. Let E be the given field and E_λ a typical countably generated subfield (always over k), then there is a 2-generator field L_λ containing E_λ, by the theorem. Let M_λ be the field coproduct of E and L_λ over E_λ; if we do this for each countably generated subfield of E we get a family $\{M_\lambda\}$ of fields, all containing E. Form their field coproduct E' amalgamating E, then in E' every countably generated subfield of E is contained in a 2-generator subfield of E', namely E_λ is contained in L_λ. Now repeat the process that led from E to E':

$$E \subset E' \subset E'' \subset \ldots \subset E^\omega \subset E^{\omega+1} \subset \ldots \subset E^\nu,$$

where $E^\alpha = \cup\{E^\beta | \beta < \alpha\}$ at a limit ordinal α, and where ν

is the first uncountable ordinal. Then E^ν is a field in which every countable subfield is contained in some E^α $(\alpha < \nu)$ and hence in some 2-generator subfield of $E^{\alpha+1} \subseteq E^\nu$. ■

At this point it is natural to ask whether there is a countable field, or one countably generated over k, containing a copy of every countable field (of a given characteristic). As in the case of groups, the answer is 'no'; this is shown by the following argument, for which I am indebted to A. Macintyre.

For any field K, let $\mathbf{S}(K)$ be the set of isomorphism types of finitely generated subgroups of K^*. If K is countable, then so is $\mathbf{S}(K)$. Now D.B. Smith [70] has shown that there are $\mathfrak{c} = 2^{\aleph_0}$ isomorphism types of finitely generated orderable groups. Further, every ordered group can be embedded in a field of prescribed characteristic, by the methods of 2.1; hence every countable ordered group can be embedded in a countable field. It follows that there are $\mathfrak{c}$ distinct sets $\mathbf{S}(K)$ as K runs over all countable fields of any given characteristic. Therefore these fields cannot all be embedded in a 2-generator field.

5.6 Extensions with different left and right degrees

In Ch.3 we examined a particular kind of binomial extension. Given a prime number p and a primitive pth root of 1, ω say, in our ground field k, let us take a field E with an endomorphism S and an S-derivation D such that

$$(1) \qquad DS = \omega SD$$

and construct fields K and L as in Th.3.4.4. We then have an extension L/K of right degree p and its left degree will be > p if we can show that $K^S \neq K$. More generally, if $[K:K^S]_L = \infty$, then $[L:K]_L = \infty$. But some care is needed here:

it is *not* enough to take $[E:E^S]_L = \infty$, for whatever S is, we shall have $K^S = K$ if $D = 0$, (because S is then an inner automorphism of L, with inverse $c \longmapsto tct^{-1}$). Likewise one can show that $[L:K]_L = [L:K]_R$ whenever K is commutative (Jacobson [56],Ch.6).

To obtain the required example, let p be a prime number, k any commutative field containing a primitive root of 1, ω say, and let Λ be any set. When k has characteristic p this is taken to mean that $\omega = 1$. We form the free algebra $F = k\langle x_{\lambda ij}\rangle$ on a family of indeterminates indexed by $\Lambda \times \mathbf{N}^2$. Let $E = \mathbf{F}(F)$ be the universal field of fractions of F. On F we have an endomorphism S defined by

$$x_{\lambda ij}^S = \omega^i x_{\lambda i\ j+1}.$$

This is an honest endomorphism because F^S is a retract of F: Let T be the endomorphism defined by

$$x_{\lambda ij}^T = \begin{cases} \omega^{-i} x_{\lambda i\ j-1} & \text{if } j > 1, \\ 0 & \text{if } j = 1. \end{cases}$$

Then $ST = 1$ (read from left to right), hence if A^S is non-full, then so is $A^{ST} = A$, i.e. S is honest, as claimed. It follows that S extends to an endomorphism of E, again denoted by S.

Next let D be the S-derivation of F defined by

$$x_{\lambda ij}^D = x_{\lambda\ i+1\ j}.$$

This again extends to an S-derivation of E, still denoted by D. We now form $L = E(t;S,D)$ and $K = E(t^p;S^p,D^p)$, as in Th.3.4.4, then L/K is a binomial extension of right degree p. To show that the left degree is > p it is enough to prove that $K^S \neq K$. This can be shown quite easily whatever Λ, e.g. we could take Λ to consist of one element. But we are then left with the task of finding whether $[K:K^S]_L$ is finite or infinite. It is almost certainly infinite, but this is not easy to show when $|\Lambda| = 1$, whereas it becomes easy for infinite Λ.

For any $\mu \in \Lambda$ denote by E_μ the subfield of E generated over k by all $x_{\lambda ij}$ such that $j > 1$, or $j = 1$ and $\lambda \neq \mu$; thus we take all x's except $x_{\mu i1}$ $(i \in N)$. It follows that $E_\mu \supseteq E^S$ for all μ, and $E_\mu(t) \supseteq L^S \supset K^S$. We claim that $x_{\lambda 11} \in E_\mu(t)$ if and only if $\lambda \neq \mu$. Assuming this for the moment, we see that the $x_{\lambda 11}$ are left linearly independent over K^S, for if $\Sigma\alpha_\lambda x_{\lambda 11} = 0 (\alpha_\lambda \in K^S)$, and some $\alpha_\mu \neq 0$, then we could express $x_{\mu 11}$ in terms of the $x_{\lambda 11}$, $\lambda \neq \mu$, and so $x_{\mu 11} \in E_\mu(t)$, which contradicts our assumption.

That $x_{\lambda 11} \in E_\mu(t)$ for $\lambda \neq \mu$ is clear from the definition. To show that $x_{\mu 11} \notin E_\mu(t)$, write $R = E_\mu[t]$ (for a fixed μ) and observe that for any $\alpha \in E$, $\alpha t = t\alpha^S + \alpha^D$, hence

$$\alpha t \equiv \alpha^D \qquad \text{(mod R)},$$

and so by induction on n, $\alpha t^n \equiv \alpha^{D^n}$ (mod R).

If $x_{\mu 11} \in E_\mu(t)$, we would have $x_{\mu 11} = fg^{-1}$, where $f,g \in R$. Then $x_{\mu 11}g \equiv f \equiv 0$ (mod R), and if $g = \Sigma t^i\beta_i$, where $\beta_i \in E_\mu$, then

$$0 \equiv \Sigma x_{\mu 11}t^i\beta_i \equiv \Sigma x_{\mu 11}^{D^i}\beta_i \quad \text{(mod R)}$$

$$\equiv \Sigma x_{\mu\, i+1\, 1}\beta_i \quad \text{(mod R)}.$$

Here we have multiplied a congruence mod R by elements of R, which is permissible. Thus we have

$$\Sigma x_{\mu\, i+1\, 1}\beta_i + \beta = 0 \qquad \beta_i, \beta \in R.$$

This is an equation in $E[t]$, more precisely in the subring $E_\mu[t] \sqcup k\langle x_{\mu i1}\rangle$ (clearly this subring is a coproduct) and by equating cofactors of the $x_{\mu i1}$ we see that $\beta = \beta_i = 0$, i.e. $g = 0$. But this is a contradiction, for g as denominator of $x_{\mu 11}$ cannot vanish. This proves that $x_{\mu 11} \notin E_\mu(t)$ and it follows that $[L:K]_L \geq |\Lambda|$.

Given any infinite cardinal α, take a set Λ of cardinal α and k a countable field, then $|L| = \alpha$, hence $[L:K]_L = \alpha$ in this case. Thus we have found an extension with right degree p and left degree α. If instead of p we have a composite integer n, pick a prime factor p of n and combine an extension of (left and right) degree n/p with the previous case. Similarly if β is an infinite cardinal $< \alpha$, we can start with an extension of degree β. Thus we have proved

Theorem 5.6.1. *Given any two cardinals* α,β *of which at least one is infinite, there is a field extension* L/K *of left degree* α *and right degree* β, *and of prescribed characteristic.* ■

Whether the left and right degrees can be both finite and different remains open. On the face of it this looks unlikely, but it does not seem an easy problem to decide.

6 · General skew field extensions

6.1 Presentations of skew fields

We have already discussed skew field extensions (in Ch.3), but they were usually of a rather special sort, of finite degree (at least on one side). We now turn to general extensions. Of course it is no longer true, as in the commutative case, that every simple extension of infinite degree is free, in fact we shall need to define what we mean by a free extension.

To reach the appropriate definition, consider a finitely generated field extension $E = K(\alpha_1,\ldots,\alpha_n)$. As before we shall take all our fields to be k-algebras, where k is a commutative field. This represents no loss of generality (in fact a gain): if k is not present we can take the prime subfield to play the role of the ground field. Given E as above, we have a homomorphism of K-rings:

$$(1) \qquad K_k\langle X\rangle \longrightarrow E, \qquad X = (x_1,\ldots,x_n), \qquad x_i \longmapsto \alpha_i.$$

Here $K_k\langle X\rangle$ is the coproduct $K \sqcup_k k\langle x_1,\ldots,x_n\rangle$; it is called the *tensor* K-*ring* on X over k. Let $\mathcal{P}$ be the singular kernel of (1); since E is an epic $K_k\langle X\rangle$-field (being generated by the α_i over K), it is determined up to isomorphism by $\mathcal{P}$. Let M be a set of matrices generating $\mathcal{P}$ as matrix ideal, then E is already determined by M; we write

$$(2) \qquad E = K_k \langle\!\langle X;M\rangle\!\rangle$$

and call this a *presentation* of E. In particular we call the α_i *free* over K if the presentation can be chosen with $M = \emptyset$; this just means that (1) is an honest map, i.e. that $\mathcal{P}$ consists of all non-full matrices and no others. From Cohn [71"] Ch.2 we know that $K_k\langle X\rangle$ is a fir and so has a universal field of fractions, written $K_k \,\langle\!\langle X \rangle\!\rangle$ and called the *free* K-*field* on X.

Given any set X and any set M of matrices over $K_k\langle X\rangle$, we can ask: When does there exist a field with presentation

(3) $\quad K_k\langle\!\langle X;M\rangle\!\rangle$?

Let (M) be the matrix ideal of $K_k\langle X\rangle$ generated by M, then there are two possibilities:

(i) (M) is improper. Then there is no field (3), in fact there is no field over which all the matrices of M become singular. Here there is no solution because we do not allow the 1-element set as a field.

(ii) (M) is proper. Now there is always a field over which the matrices of M become singular, possibly more than one. The different such fields correspond to the prime matrix ideals containing (M), and there is a universal one among them precisely when the radical $\sqrt{(M)}$ is prime. In particular, this is so when (M) is prime, and that will be the only case in which the notation (3) will be used.

Let E be a field with presentation (3); we shall say that E is *finitely related* when M can be chosen finite, and E is *finitely presented* if X and M can both be chosen finite. As for groups we have

Theorem 6.1.1 *A finitely related field can be expressed*

as the field coproduct of a fintely presented and a free field.

Proof. Let $E = K_k\langle X;M\rangle$, where M is finite. Then the set X' of elements of X occurring in matrices from M is finite. Let X'' be the complement of X' in X, then we clearly have $E = E' \circ E''$, where $E' = K_k\langle X';M\rangle$, $E'' = K_k\langle X''\rangle$. Here E' is finitely presented and E'' is free.■

In the special case when E/K has finite degree, the above construction can be a little simplified. In that case (1) is surjective, not merely epic. Moreover, instead of taking the free algebra, we can incorporate the commutativity relations as follows. Let $u_1,\ldots,u_n$ be a right K-basis of E, then since E is a K-bimodule, we have the equations

$$(4) \qquad \alpha u_j = \Sigma u_i \rho_{ij}(\alpha) \qquad\qquad (\alpha \in K),$$

where $\alpha \longmapsto (\rho_{ij}(\alpha))$ is a homomorphism from K to K_n. Let M be the free right K-space on $u_1,\ldots,u_n$ as basis; by the equations (4), M becomes a K-bimodule, which contains K as submodule (as we see by choosing our basis of E so that $u_1 = 1$). Let $\Phi_K(M)$ be the filtered ring on this bimodule, constructed as in 2.5 of Cohn [71'']; by Th.2.5.1, l.c., $\Phi_K(M)$ has weak algorithm and hence is a fir. Now E is obtained as a homomorphic image of $\Phi_K(M)$; so we need to look for ideals in $\Phi_K(M)$ which as right K-spaces have finite codimension, in fact the kernel of (1) in this case is a complement of M in $\Phi_K(M)$. But it is not at all clear how this would help in the classification of extensions of finite degree.

In practice most of the presentations we shall meet are given by equations rather than singularities, but the latter are important in theoretical considerations, e.g. when we want to prove that an extension is free we must check that there are no matrix singularities. We shall return to this question in 8.1.

A special case occurs when E (and hence also K) is of finite degree over k. In that case the singularity of a matrix can be expressed by the vanishing of a norm and hence it will be enough to consider equations. Only in the case of infinite extensions are the matrix singularities really needed.

6.2 Existentially closed skew fields

Let k be a commutative field. By an *algebraic closure* of k one usually understands a field $\bar{k}$ with the properties:

(i) $\bar{k}$ is algebraic over k,

(ii) every equation over $\bar{k}$ has a solution in $\bar{k}$.

It is well known that every commutative field has an algebraic closure, and that the latter is unique up to isomorphism (though not necessarily a unique isomorphism, thus the algebraic closure is not a functor). When one tries to perform an analogous construction for skew fields one soon finds that it is impossible to combine (i) and (ii). In fact (i) is rather restrictive, so we give it up altogether and concentrate on (ii). Here it is convenient to separate two problems, namely (a) which equations are soluble (in some extension) and (b) whether every soluble equation has a solution in the closure. Of these (a) is a difficult question to which we shall return later, and for the moment concentrate on (b). The assertion that an equation $f(x_1,\ldots,x$ $=0$ has a solution can be expressed as follows:

$$\exists a_1,\ldots,a_n \; f(a_1,\ldots,a_n) = 0.$$

Any sentence of the form $\exists a_1,\ldots,a_n \; P(a_1,\ldots,a_n)$, where P is an expression obtained from equations by negation, conjunction and disjunction is called an *existential sentence*.

By an *existentially closed* field, EC-*field* for short, we understand a field K (over k) such that any existential sentence which holds in some field extension of K, already holds in K. Clearly such a sentence can always be expressed as a finite conjunction of disjunctions of basic formulae, a basic formula being of the form $f = g$ or its negation, where f,g are polynomials in $x_1,\ldots,x_r$, i.e. elements of $K_k\langle x_1,\ldots, x_r\rangle$.

Now the negation of $f = g$, i.e. $f \neq g$, can again be expressed as an equation, viz. $(f - g)y = 1$, where y is a new variable, and any disjunction of equations can be expressed as a single equation, since $(f_1 = g_1)\vee \ldots \vee(f_n = g_n)$ holds if and only if $(f_1 - g_1) \ldots (f_n - g_n) = 0$. So we have reduced the sentence to a finite set of equations, and we see that K is existentially closed if and only if any finite system of equations which is consistent (i.e. has a solution in some extension field) has a solution in K itself. For example, k itself is existentially closed over k precisely when k is algebraically closed, but for $K \supset k$ it may be possible for K to be existentially closed even when k is not algebraically closed. In fact we shall see that every field K can be embedded in an EC-field, but the latter will not be unique in any way.

Instead of the vanishing of elements, i.e. equations, we may equally well talk about the singularity of matrices. For if $A = (a_{ij})$ is any n x n matrix, let us write sing(A) for the existential sentence

$$\exists u_1,\ldots,u_n,v_1,\ldots,v_n(\Sigma a_{1j}u_j = 0 \wedge \ldots \wedge \Sigma a_{nj}u_j = 0 \wedge$$
$$(1 - u_1v_1)\ldots(1 - u_nv_n) = 0),$$

and nonsing(A) for the sentence

$$\exists x_{ij} (i,j = 1,\ldots,n)(\Sigma a_{i\nu} x_{\nu j} = \delta_{ij} \; (i,j = 1,\ldots,n)).$$

It is clear that sing(A) asserts that (over a field) A is singular and

$$\text{nonsing}(A) \Longleftrightarrow \neg\, \text{sing}(A),$$

(where $\neg P$ means 'not P'). From this it is easy to deduce

Proposition 6.2.1. *A field* K *over* k *is an* EC-*field if and only if any finite set of matrices over* $K_k\langle X\rangle$ *which all become singular for a certain set of values of the* x*'s in some extension of* K*, already become singular for some set of values in* K.

Proof. The condition for existential closure concerns the vanishing of a finite set of elements, i.e. the singularity of 1 x 1 matrices, and so is a special case of the second condition, which is therefore sufficient. Conversely, when K is existentially closed, and we are given matrices A_1, $\ldots,A_r$ which become singular in some extension, then sing $(A_1) \wedge \ldots \wedge \text{sing}(A_r)$ is consistent and hence has a solution in K. ■

It is almost trivial to show that every field K can be embedded in an EC-field, by first constructing an extension K_1 in which a given finite consistent system of equations over K has a solution, and then repeating the process infinitely often. However there is no guarantee that the EC-field so obtained will contain solutions of every finite consistent system over K. For this to hold we need to be assured that any two consistent systems over K are jointly consistent. Of course this follows from the existence of field coproducts: if Φ_1, Φ_2 are two consistent systems of equations over K, say Φ_i has a solution in K_i, then any field L containing both K_1 and K_2 will contain a solution of $\Phi_1 \wedge \Phi_2$. For L we can take e.g. the field coproduct

$K_1 \circ K_2$; more generally a class of algebras is said to possess the *amalgamation property* if any two extensions B_1, B_2 of an algebra A are contained in some algebra C. An example of a class *not* possessing the amalgamation property is the class of formally real fields.

To construct an EC-field extension of K we take the family $\{C_\lambda\}$ of all finite consistent systems of equations over K and for each λ take an extension E_λ in which C_λ has a solution. Put $K_1 = \overset{\circ}{K} E_\lambda$, then every finite consistent set of equations over K has a solution in K_1. If we repeat this process, we obtain a tower

$$K \subset K_1 \subset K_2 \subset \ldots$$

whose union L is again a field, of the same cardinal as K or countable (if K was finite). Any finite consistent set of equations over L has its coefficients in some K_i and so has a solution in K_{i+1}. Thus L is an EC-field and we have proved

Theorem 6.2.2. *Let* K *be any field (over* k*), then there exists an* EC*-field* L *containing* K*, in which every finite consistent set of equations over* K *has a solution. When* K *is infinite,* L *can be chosen to have the same cardinal as* K*, while for finite* K*,* L *may be taken countable.* ∎

If α is any infinite cardinal, we can similarly construct α-EC-fields containing a given field K, in which every consistent set of fewer than α equations has a solution. However, the EC-fields constructed here are not in any way unique; even a minimal EC-field containing a given field K need not be unique up to isomorphism, as will become clear later on. Further, it will no longer be possible to find an EC-field algebraic (in any sense) over K.

Sometimes a stronger version of algebraic closure is needed, in which the above property holds for *all* sentences, not merely existential ones. We shall not need this stronger form, and therefore merely state the results without proof.

Let A be an *inductive* class of algebras (of some sort), i.e. a class closed under isomorphisms and unions of chains. By an *infinite forcing companion* one understands a subclass C of A such that

F.1 *Every* A-*algebra can be embedded in a* C-*algebra,*

F.2 *Any inclusion* $C_1 \subseteq C_2$ *between* C-*algebras is elementary,*

F.3 C *is maximal subject to* F.1,2.

(Recall that a map $f:P \longrightarrow Q$ is *elementary* if for every sentence $\alpha(x)$ which holds in P, $\alpha(xf)$ holds in Q.) It can be shown that every inductive class has a unique forcing companion (cf. A. Robinson [71], G. Cherlin [72], J. Hirschfeld-W.H. Wheeler [75]). All this applies in particular to skew fields. Here we also have the amalgamation property; but an important difference between commutative and non-commutative fields is that algebraically closed commutative fields are axiomatisable (we can write down a set of first order sentences asserting that all equations have solutions); the corresponding statement for EC-fields is false. This follows from the fact that the class of EC-fields is not closed under ultrapowers (J. Hirschfeld-W.H. Wheeler [75]).

Although EC-fields do not share all the good properties of algebraically closed fields, they have certain new features not present in the commutative case. For example, the property of being transcendental over the ground field can now be expressed as an elementary sentence:

$$(1) \qquad \text{transc}(x): \quad \exists y,z(xy = yx^2 \wedge x^2z = zx^2 \wedge xz \neq zx \wedge y \neq$$

This sentence is due to Wheeler (l.c.). It states that there is an element z commuting with x^2 but not with x, hence $k(x^2) \subset k(x)$; secondly x is conjugate to x^2, so $k(x) \cong k(x^2)$, in particular, $[k(x):k] = [k(x^2):k]$, and so, because $k(x^2) \subset k(x)$, the degree must be infinite. Conversely, when x is transcendental, (1) can be satisfied in some extension, and hence in the EC-field.

Another sentence characterizing transcendental elements, in the case of a perfect ground field, was obtained by Boffa-v.Praag [72]; it is

$$\exists y(xy - yx = 1).$$

Wheeler has generalized (1) to find (for each $n \geq 1$) an elementary formula $\mathrm{transc}_n(x_1,\ldots,x_n)$, expressing the fact that $x_1,\ldots,x_n$ commute pairwise and are algebraically independent over the ground field; as a consequence he is able to show that every EC-field contains a commutative algebraically closed subfield of infinite transcendence degree.

To describe EC-fields in a little more detail we need two results from Hirschfeld-Wheeler [75], see also Macintyre [75].

Lemma 6.2.3 *(Zig-zag lemma) If* K, L *are two countable* EC-*fields over* k, *then* $K \cong L$ *if and only if they have the same family of finitely generated subfields.*

Proof. Clearly the condition is necessary. Conversely, let K,L be countable EC-fields having the same finitely generated subfields. Let $K = k(a_1,a_2,\ldots)$, $L = k(b_1,b_2,\ldots)$; we shall construct finitely generated subfields K_n, L_n of K and L respectively such that (i) $K_n \subseteq K_{n+1}$, $L_n \subseteq L_{n+1}$, (ii) $K_n \supseteq k(a_1,\ldots,a_n)$, $L_n \supseteq k(b_1,\ldots,b_n)$, (iii) there is an isomorphism between K_{n+1} and L_{n+1} extending a given isomorphism between K_n and L_n. - Since $K = \cup K_n$, $L = \cup L_n$, it will follow that $K \cong L$, by taking the common extension in (iii).

Put $K_o = L_o = k$; if K_n, L_n are defined, with an isomorphism $\phi_n : K_n \longrightarrow L_n$, let $K'_n = K_n(a_{n+1})$, then K'_n is finitely generated, hence isomorphic to a subfield of L containing an isomorphic copy of L_n. By Th.5.5.1, Cor.1, L can be embedded in a finitely homogeneous extension, but L is an EC-field and hence is itself finitely homogeneous. Thus we can apply an inner automorphism of L so as to map K'_n onto a subfield L'_n containing L_n, in such a way that the restriction to K_n is the homomorphism ϕ_n. Let $\phi'_n : K'_n \longrightarrow L'_n$ be the isomorphism so obtained. Now put $L_{n+1} = L'_n(b_{n+1})$ and find an isomorphic copy of L_{n+1} in K; this will contain a subfield isomorphic to K'_n and by applying a suitable inner automorphism of K we obtain an isomorphism of L_{n+1} with a subfield K_{n+1} say of K, which when restricted to L'_n is $(\phi'_n)^{-1}$. Now K_{n+1}, L_{n+1} satisfy (i)-(iii) and the result follows by induction.■

For the second result let us write, for any subset S of K, $C(S)$ for the centralizer of S in K:

Proposition 6.2.4. *Let* K *be an* EC-*field over* k *and let* $a_1,\ldots,a_r,b \in K$, *then*

$$b \in k(a_1,\ldots,a_r) \iff C(b) \supseteq C(a_1,\ldots,a_r).$$

This means that the formula '$b \in k(a_1,\ldots a_r)$', not at first sight elementary (and in fact not so in the commutative case), can be expressed as an elementary sentence in an EC-field:

$$(2) \qquad \forall x \quad (a_i x = x a_i (i = 1,\ldots,r) \Longrightarrow bx = xb).$$

Proof. Write $A = k(a_1,\ldots,a_r)$, then if $b \in A$, (2) clearly holds; if $b \notin A$, then (2) is false in $K \overset{\circ}{_A} A(x)$ and hence also in K, because the latter is existentially closed.■

Taking r = 0, we obtain a result which is well known in

the special case when k is the prime subfield:

Corollary 1. *The centre of an* EC-*field over* k *is* k. ■

EC-fields are in some way analogous to algebraically closed groups, which have been studied by B.H. Neumann [73]; the next result is analogous to a property proved by Neumann for groups:

Proposition 6.2.5 *An* EC-*field cannot be finitely generated or finitely related.*

Proof. Given $a_1,\ldots,a_n \in K$, the sentence

$$\exists x,y(a_i x = xa_i (i = 1,\ldots,n) \wedge xy \neq yx)$$

is consistent, for it holds in $K(x) \circ_k k(y)$; hence it holds in K itself, and by Prop. 6.2.3 this means that K contains an element $y \notin k(a_1,\ldots,a_n)$. Hence K cannot be finitely generated. In a finitely related field which is not finitely generated, infinitely many generators occur in no relation and so generate a free factor. If x is one of them, then the sentence $\exists y(x = y^2)$ is not satisfied in K, though clearly consistent, and this contradicts the fact that K is an EC-field. Hence K cannot be finitely related. ■

As we saw in 5.5, there are continuum-many non-isomorphic finitely generated fields, hence no countable EC-field can contain them all, i.e. there are no countable *universal* EC-fields (Hirschfeld-Wheeler [75]). However, it is possible to construct a countable EC-field containing all finitely presented fields: we simply enumerate all finitely presented fields $K_1,K_2,\ldots$ over k, form their field product over k and take a countable EC-field containing this product, which exists by Th.6.2.2. The result is a countable EC-field containing each finitely presented field over k.

Any EC-field has proper EC-subfields, thus there are no minimal EC-fields. This follows from

Theorem 6.2.6. *Let* K *be an* EC-*field over* k *and* c *any element*

of K, *then the centralizer* C *of* c *in* K *is an* EC-*field over* k(c).

Proof. It is clear that k(c) is contained in the centre of C. Now let

$$(3) \quad f_1 = \ldots = f_r = 0$$

be any finite set of equations in $x_1, \ldots, x_n$ over C which has a solution in some extension of C over k(c). This means that the solution also satisfies

$$(4) \quad x_1 c - c x_1 = \ldots = x_n c - c x_n = 0.$$

Hence the equations (3), (4) are consistent and so have a solution in K. By (4) this means that we have found a solution of (3) in C, so C is an EC-field over k(c), as claimed.

By taking intersections we get EC-fields over k itself. Such a construction can also be obtained in a more straightforward fashion:

Theorem 6.2.7. *Let* K *be an* EC-*field over* k *and let* a ε K *be transcendental over* k, *then there exists* b ε K *such that*

$$(5) \quad ba = a^2 b \neq 0.$$

If C *is the centralizer in* K *of such a pair* a,b, *then* C *is again an* EC-*field over* k, *and the inclusion* C $\subseteq$ K *is an elementary embedding*.

Proof. Since a is transcendental over k, the mapping $f(a) \longmapsto f(a^2)$ is an endomorphism σ of k(a), so the system (5) has a solution in some extension of K, and hence in K itself. Now given a,b satisfying (5), let C be their centralizer in K and let

$$(6) \quad f_1 = \ldots = f_r = 0$$

be a consistent system of equations in the variables $x_1, \ldots, x_n$ over C. Since K is an EC-field, this system has a solution in K. Let $c_1, \ldots, c_s \in C$ be the coefficients occurring in (6) and consider the system consisting of (6) and

$$(7) \quad x_i y = y x_i, \quad x_i z = z x_i, \quad c_j y = y c_j, \quad c_j z = z c_j, \quad zy = y^2 z \neq 0.$$

This system is consistent: we form first K(y) and with the endomorphism $\sigma : f(y) \longmapsto f(y^2)$ form $K(y)(z;\sigma)$. Hence (7) has a solution in K itself; let us denote this solution also by x_i, y, z. Then the mapping $y \longmapsto a$, $z \longmapsto b$ defines an isomorphism

$$k(c_1, \ldots, c_s, y, z) \cong k(c_1, \ldots, c_s, a, b),$$

for both sides are obtained by first adjoining a central indeterminate y and then forming the field of fractions of the skew polynomial ring with respect to the endomorphism $f(y) \longmapsto f(y^2)$. By homogeneity there exists $t \in K$ such that $t^{-1} c_j t = c_j$, $t^{-1} y t = a$, $t^{-1} z t = b$. Now put $t^{-1} x_i t = x_i'$, then $x_i' \in C$ and x_i' is a solution of (6). This shows C to be an EC-field.

To prove that the inclusion $C \subseteq K$ is an elementary embedding we need only show that every finitely generated subfield of K can be embedded in C. Let $c_1, \ldots, c_s \in K$ and consider the system (7), but without the equations involving x_i. This system is consistent and so has a solution in K. Since $k(y,z) \cong k(a,b)$ with $y \longmapsto a$, $z \longmapsto b$, there exists $t \in K$ such that $yt = ta \neq 0$, $zt = tb$. If we put $t^{-1} c_j t = c_j'$, then $c_j' \in C$ and $k(c_1, \ldots, c_s) \cong k(c_1', \ldots, c_s')$, hence the result. ■

When K is countable, it follows from the zig-zag lemma (6.2.3) that $C \cong K$ and we obtain the

Corollary. *Every countable* EC-*field has a proper subfield*

isomorphic to itself. ■

An important and useful result due to Wheeler (l.c.) is that every countable EC-field has outer automorphisms; the proof below is taken from Cohn [75].

Theorem 6.2.8. *Every countable* EC-*field has* $2^{\aleph_0}$ *distinct automorphisms, and hence has outer automorphisms.*

Proof. Let K be generated over k by $a_1, a_2, \ldots$, where the a_i are chosen so that $a_n \notin k(a_1, \ldots, a_{n-1})$; this is clearly possible. By Prop. 6.2.4 there exists b_n commuting with $a_1, \ldots, a_{n-1}$ but not with a_n. Let β_n be the inner automorphism induced by b_n and consider the formal product

$$\alpha = \beta_1^{\varepsilon_1} \beta_2^{\varepsilon_2} \ldots,$$

for a given choice of exponents $\varepsilon_i = 0,1$. We claim that α defines an automorphism on K. Its effect on $k(a_1, \ldots, a_n)$ is

$$\beta_1^{\varepsilon_1} \ldots \beta_n^{\varepsilon_n},$$

for when $i > n$, β_i leaves $k(a_1, \ldots a_n)$ elementwise fixed. Thus it is an endomorphism which is in fact invertible since each β_i is. Since the ε_i are independent and each choice gives a different automorphism, we have indeed $2^{\aleph_0}$ distinct automorphisms; of course there cannot be more than this number. Now a countable field has at most countably many inner automorphisms, hence K has outer automorphisms. ■

This proof is of course highly non-constructive; since EC-fields themselves are not given in any very explicit form, there seems little hope of actually finding a particular outer automorphism.

An important but difficult question is: Which fields are embeddable in finitely presented fields? It would be interesting if some analogue of Higman's theorem could be established. This asserts that a finitely generated group is embeddable in a finitely presented group if and only if it is recursively presented (Higman [61]).

6.3 A specialization lemma

In this section we digress somewhat to prove a technical result which is sometimes useful:

Lemma 6.3.1 *(Specialization lemma) Let* K *be a field with centre* C *and assume* (i) C *is infinite and* (ii) K *has infinite degree over* C. *Then any full matrix over* $K_C\langle X\rangle$ *is non-singular for some set of values of* X *in* K.

Some preparations are necessary for the proof. In the first place we shall need Amitsur's theorem on generalized polynomial identities. Let A be a k-algebra, then by a *generalized polynomial identity* (g.p.i.) one understands a non-zero element p of $A_k\langle X\rangle$ which vanishes under all mappings $X \longrightarrow A$. Amitsur [65] proved that a primitive k-algebra A satisfies a g.p.i. if and only if it is a dense ring of linear transformations over a skew field of finite degree over its centre and A contains a transformation of finite rank. We shall be particularly concerned with the case where A is itself a skew field; in this case Amitsur's theorem takes the following form:

A skew field satisfies a generalized polynomial identity if and only if it is of finite degree over its centre.

For the proof we refer to Amitsur [65].

A second result is the inertia theorem (Bergman [67], Cohn [71"]). Let R be any ring and A a subring, then A is said to be n-*inert* in R if for any families (a_λ) of rows in R^n and (b_μ) of columns in ${}^nR$ such that $a_\lambda b_\mu \in A$ for all λ, μ, there exists $P \in GL_n(R)$ such that on writing $a'_\lambda = a_\lambda P$, $b'_\mu = P^{-1}b_\mu$, each product $a'_\lambda b'_\mu$ lies trivially in A, in the sense that for each $i = 1,\ldots,n$, either $a'_{\lambda i} = 0$ or $b'_{\mu i} = 0$ or both $a'_{\lambda i}$ and $b'_{\mu i}$ lie in A. If A is n-inert in R for all n, it is called *totally inert* in R. Now we have the

Inertia theorem. $K_k\langle X\rangle$ *is totally inert in* $K_k\langle\langle X\rangle\rangle$.

The theorem is proved in Cohn [71"] (p.103f.) for a wider class of rings; however the proof given there is not complete. We therefore give a proof below (which it is hoped is complete). In the proof we shall need the weak algorithm; for this we refer to Cohn [71"], and in fact, the reader willing to accept the inertia theorem will need only the following corollary in which the notion of inertia does not appear.

Corollary (to the inertia theorem). The embedding $K_k\langle X\rangle \longrightarrow K_k\langle\langle X\rangle\rangle$ *is honest.*

For let C be a full matrix over $F = K_k\langle X\rangle$ which is non-full over $\hat{F} = K_k\langle\langle X\rangle\rangle$, say $C = AB$, where A is $n \times r$, B is $r \times n$ and $r < n$. By inertia we can find $P \in GL_r(\hat{F})$ such that on writing $A' = AP$, $B' = P^{-1}B$, the product of any row of A' by any column of B' lies trivially in F. Since C is full, A', B' cannot have all their entries in F. If the (1,1)-entry of A is not in F, say, then the first row of B' is zero, and on omitting the first column of A' and the first row of B' we can diminish r. By induction on r we obtain a contradiction; this proves the corollary, starting from the theorem. ■

We shall prove the inertia theorem in the following (slightly more general) form. The proof follows Cohn-Dicks [76].

Inertia theorem. Let R *be a graded ring with weak algorithm, and* $\hat{R}$ *its completion, then* R *is totally inert in* $\hat{R}$.

Proof. For any $a \in \hat{R}$ the *order* o(a) of a is defined as the minimum of the degrees of the homogeneous components of a, or ∞ if a=0. Let

$$\mathfrak{m} = \{x \in R \mid o(x) > 0\},$$

then by the weak algorithm $R/\mathfrak{m}$ is a field; moreover $\hat{R}$ is the completion of R in the $\mathfrak{m}$-adic topology. We shall write $\hat{\mathfrak{m}}$ for the completion of $\mathfrak{m}$. Now $\mathfrak{m}$ as a free right ideal of R has, by the weak algorithm, a homogeneous basis X say, and any $a \in \hat{\mathfrak{m}}$ can be uniquely written as $a = \Sigma x a_x$, where the summation is over all $x \in X$ and all but a finite number of the $a_x \in \hat{R}$ are zero. We note that

$$(1) \quad o(a) \geq 1 + \min_X\{o(a_x)\};$$

further, if $a \in \mathfrak{m}$, then all the a_x lie in R.

Let $A \subseteq \hat{R}^r$, $B \subseteq {}^r\hat{R}$ be such that $ab = \Sigma a_i b_i \in R$ for all $a \in A$, $b \in B$. We may assume that A, B are closed in the sense that each consists of all the rows (resp. columns) which transport the columns (resp.rows) of the other into R. Then A will be a left R-module and B a right R-module. We claim that A has the following property:

$$(2) \quad A \cap (\hat{\mathfrak{m}})^r = \mathfrak{m}A,$$

i.e. if the components of $a \in A$ have positive order, then $a = \Sigma x a_x$, where $a_x \in A$.

Clearly $a = \Sigma x a_x$ for a unique $a_x \in \hat{R}^r$; we have to show that $a_x \in A$. Now for all $b \in B$, $\Sigma x(a_x b) = ab \in R$, hence $a_x b \in R$ and by our closure assumption $a_x \in A$, i.e. (2).

Consider the image of A in $(\hat{R}/\hat{\mathfrak{m}})^r$, which by (2) is just $A/\mathfrak{m}A$. It is a left $\hat{R}/\hat{\mathfrak{m}} = R/\mathfrak{m}$-space of dimension s say, where $s \leq r$. By $\hat{R}$-column operations on A (and the corresponding row operations on B) we may assume that A contains $e_1, \ldots, e_s$, part of the standard basis for row vectors, while any component after the first s in any element of A is a non-unit, i.e. has positive order.

Consider the case $s < r$. We claim that for all $a =$

$(a_1,\ldots,a_r) \in A$ we have $a_r = 0$. If not, choose $a \in A$ so as to minimize $o(a_r)$. By adding left multiples of $e_1,\ldots,$ e_s to a we may assume that each component has positive order. Then by (2), $a = \Sigma x a_x$, $a_x \in A$ and by (1) the rth coordinate of some a_x must have lower order than a_r, which contradicts the minimality of $o(a_r)$. Hence $a_r = 0$ and this effectively reduces r, so we may complete the proof by induction on r.

There remains the case $s = r$. Then $A \supseteq R^r$, hence $B \subseteq {}^rR$, and by symmetry we may assume that the $R/\mathfrak{m}$-dimension of $B/B\mathfrak{m}$ is also r. Now R is a fir, so B is free as right R-module on a set W say. Hence $B \otimes_R R/\mathfrak{m} = B/B\mathfrak{m}$ is free as right $R/\mathfrak{m}$-module on the image of W. But $B/B\mathfrak{m}$ has dimension r, so W has precisely r elements, and by the dual of (2) they are left $R/\mathfrak{m}$-independent (mod $\mathfrak{m}$). Let P be the $r \times r$ matrix whose columns are the elements of W. Since P is invertible (mod $\mathfrak{m}$), it is invertible over $\hat{R}$. Now $B = WR$, so $B = P({}^rR)$, hence $P^{-1}B = {}^rR$ and $AP = R^r$, which proves the assertion. ■

For the proof of the specialization lemma we also need a couple of auxiliary results:

Lemma 6.3.2. *Let* A *be any matrix over a skew field, partitioned as*

$$A = \begin{pmatrix} A_1 & A_2 \\ A_3 & A_4 \end{pmatrix}$$

where A_1 *is an invertible* $r \times r$ *matrix. Then rank* $A \geq r$, *with equality if and only if* $A_4 = A_3A_1^{-1}A_2$ *and then*

$$A = \begin{pmatrix} A_1 \\ A_3 \end{pmatrix} (I \quad A_1^{-1}A_2).$$

Proof. By elementary transformations we can change A to the form

$$\begin{pmatrix} A_1 & A_2 \\ 0 & A_4 - A_3A_1^{-1}A_2 \end{pmatrix}.$$

Here the first r columns are linearly independent and there will be more linearly independent columns unless $A_4 = A_3A_1^{-1}A_2$ and in that case

$$A = \begin{pmatrix} A_1 & A_2 \\ A_3 & A_3A_1^{-1}A_2 \end{pmatrix} = \begin{pmatrix} A_1 \\ A_3 \end{pmatrix} (I \quad A_1^{-1}A_2) \quad . ■$$

Lemma 6.3.3. *Let* K *be a skew field with infinite centre* C *and consider the polynomial ring* K[t] *in a central indeterminate* t, *with field of fractions* K(t). *If* A = A(t) *is a square matrix over* K[t], *then the rank of* A *over* K(t) *is the supremum of the ranks of* $A(\tau)$, $\tau \in C$. *In fact this supremum is assumed for all but a finite number of values of* τ.

Proof. We transform A to diagonal form by PAQ-reduction over the principal ideal domain K[t]. This means that we can find invertible matrices P, Q over K[t] such that

$$PAQ = \mathrm{diag}(\alpha_1,\ldots,\alpha_n) \qquad (\alpha_i \in K[t]).$$

The product of the diagonal terms on the right gives us a polynomial f whose zeros in C are the only points of C at which A = A(t) falls short of its maximum rank, and the number of these points cannot exceed deg f = n. ■

We are now ready to prove the specialization lemma (Lemma 6.3.1). Let A = A(X) be an n x n matrix over $K_C\langle X\rangle$ and r the sup of the ranks as its arguments range over K^X. By a change of variables we may assume that the maximum rank is assumed at the point $0 \in K^X$, and by elementary operations we may take the principal r x r minor of A(0) to be invertible. Thus if

$$A(X) = \begin{pmatrix} B_1(X) & B_2(X) \\ B_3(X) & B_4(X) \end{pmatrix}$$

then $B_1(0)$ is invertible.

Given any $a \in K^X$ and any $\tau \in C$, we have rank $A(\tau a) \leq r$, hence by Lemma 6.3.3, the rank of $A(ta)$ over $K[t]$ is $\leq r$ and so the same holds over $K((t))$. Now $B_1(ta)$ is a polynomial in t with matrix coefficients and constant term $B_1(0)$, a unit, hence $B_1(ta)$ is a unit in $K((t))$. By Lemma 6.3.2, the equation

$$(3) \quad B_4(ta) = B_3(ta)B_1(ta)^{-1}B_2(ta)$$

holds in $K((t))$ and hence in $K[\![t]\!]$, for all $a \in K^X$. But the coefficient of every power of t in (3) is a polynomial in the components of (3), thus (3) holds identically over K and by Amitsur's theorem on generalized polynomial identities it follows that

$$(4) \quad B_4(tX) = B_3(tX)B_1(tX)^{-1}B_2(tX)$$

holds identically in the completion of $R = K_C\langle X\rangle[t]$ in the t-adic topology. Hence the same equation holds in the completion in the (t,X)-adic topology, and also in that of the X-adic topology; for in each case the matrix $B_1(tX)$ has constant term $B_1(0)$ and so is invertible. Thus we may set $t = 1$ in (4) and find that $B_4(X) = B_3(X)B_1(X)^{-1}B_2(X)$ in $K_C\langle\langle X\rangle\rangle$, i.e. A(X) is non-full over this ring; by the Corollary to the inertia theorem it follows that A(X) is non-full over $K_C\langle X\rangle$. It follows that every full matrix over $K_C\langle X\rangle$ is non-singular for some value in K. ■

For the specialization lemma to hold it is clearly necessary that C should be the precise centre of K. For if there existed α in the centre of K but not in the ground field, then $\alpha x - x\alpha$ is non-zero and hence full, even though it vanishes for all values of x in K. Secondly if $[K:C]$ is finite, there are non-trivial identities over K, so again the specialization lemma does not hold. On the other hand it is not known whether the hypothesis that C be infinite is essential. It could be omitted if the following rather plausible conjecture could be proved:

Conjecture. *Let* K *be an infinite field with a finite centre, then for any square matrix* A *over* K *there exists* $\alpha \in K$ *such that* $A - \alpha I$ *is non-singular.*

This is known to be true if the centre is infinite (cf. Lemma 6.3.3) and false when K is itself finite.

The proof of the specialization lemma (which owes several simplifications to G.M. Bergman) is taken from Cohn [72']. Some applications will be given in the next section. But first we shall deal with the promised generalization of Prop.5.4.2.

Lemma 6.3.4. *Let* E *be a field over* k *and* t *a central indeterminate, then the mapping*

$$E_k\langle X\rangle \longrightarrow E(t)_{k(t)}\langle X\rangle$$

is honest.

Proof. Let us write $F = E_k\langle X\rangle$, then there is a natural inclusion mapping $E \longrightarrow F$, hence a mapping $E(t) \longrightarrow F(t)$ and so an E(t)-ring homomorphism

$$(5) \quad E(t)_{k(t)}\langle X\rangle \longrightarrow F(t).$$

Now let A be a full matrix over $E_k\langle X\rangle$, then A is invertible over F, hence invertible over F(t) and by (5) it is full over $E(t)_{k(t)}\langle X\rangle$, as we had to show. ∎

Lemma 6.3.5. *Let* E *be a field with centre* k*, then* E(t) *has* *the centre* $k(t)$.

Proof. Every element of $E(t)$ has the form $\phi = fg^{-1}$, where $f,g \in E[t]$. We shall use induction on $d(\phi) = \deg f + \deg g$ to prove that if ϕ is in the centre of $E(t)$, then $\phi \in k(t)$. For $d(\phi) = 0$ the result holds by hypothesis. If $d(\phi) > 0$, we may assume $\deg f \geq \deg g$, replacing ϕ by ϕ^{-1} if necessary. By the Euclidean algorithm, $f = qg + r$, where $\deg r < \deg g$, with uniquely determined $q,r \in E[t]$. Let us write $u^c = c^{-1}uc$ for any $c \in E^*$, then

$$\phi = fg^{-1} = q + rg^{-1}, \qquad \phi^c = q^c + r^c(g^c)^{-1}.$$

Since ϕ is in the centre of $E(t)$, we have $q + rg^{-1} = q^c + r^c(g^c)^{-1}$, i.e.

$$q - q^c = r^c(g^c)^{-1} - rg^{-1}. \tag{6}$$

Now $v(\phi) = \deg g - \deg f$ is a valuation on $E(t)$, and the left-hand side of (6) has value ≤ 0, unless $q^c = q$, while the right-hand side has positive value. Hence both sides are 0, $q^c = q$ and rg^{-1} is in the centre, but $d(rg^{-1}) < d(fg^{-1})$, so the result follows by induction. ■

Theorem 6.3.6. *Let* F *be a field with centre* C, *let* E *be a subfield of* F *and put* $k = E \cap C$, *then there is a natural embedding*

$$E_k\langle X\rangle \longrightarrow F_C\langle X\rangle.$$

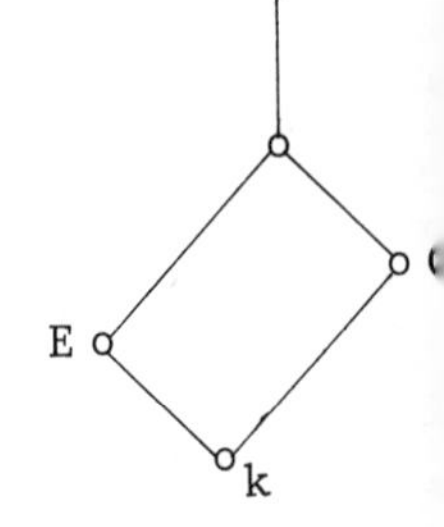

Proof. We have to show that the homomorphism (7) is honest.

$$E_k\langle X\rangle \longrightarrow F_C\langle X\rangle \tag{7}$$

Assume first that k is infinite. Take an infinite set Y and put $F' = F_C\langle Y\rangle$. Let E' be the subfield of F' generated by E and Y, then E' has precise centre k; for if $a \in E'$, then a involves only finitely many elements of Y, so we can find $y \in Y$ such that y does not occur in a. Now $ay = ya$ can hold only if a involves no element of Y, thus $a \in F$, and moreover a must be in C; it is also in E, hence $a \in C \cap E = k$.

Now consider the following diagram

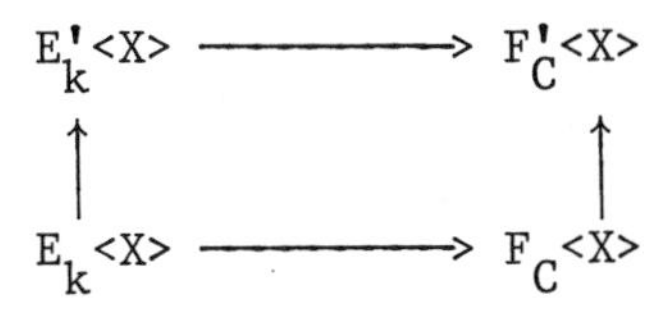

Let A be a matrix over $E_k\langle X\rangle$ which is full; by Prop. 5.4.2 it is full over $E'_k\langle X\rangle$ and so, by the specialization lemma, A becomes non-singular for some values of X in E'; hence it is full over $F'_C\langle X\rangle$ and therefore also over $F_C\langle X\rangle$. This shows (7) to be honest.

If k is finite, the above argument shows that the mapping

$$E(t)_{k(t)}\langle X\rangle \longrightarrow F(t)_{C(t)}\langle X\rangle$$

is honest, because clearly $C(t) \cap E(t) = k(t)$ and by Lemma 6.3.5, C(t) is the centre of F(t). Moreover, the mapping $E_k\langle X\rangle \longrightarrow E(t)_{k(t)}\langle X\rangle$ is honest, by Lemma 6.3.4, hence any full matrix over $E_k\langle X\rangle$ is full over $F(t)_{C(t)}\langle X\rangle$ and hence over $F_C\langle X\rangle$. ■

6.4 The word problem for free fields

The word problem in a variety of algebras, e.g. groups, is the problem of deciding, for a given presentation of a group, when two expressions represent the same group element. In the case of skew fields we again have a presentation, as explained in 6.1, and we can ask the same question, but the word problem is now a relative one. Generally we have a coefficient field K and we need to know how K is given. It may be that K itself is given by a presentation with solvable word problem, and the algorithm which achieves this is then incorporated in the algorithm to be constructed; or more generally, we merely postulate that certain questions about K can be answered in a finite number of steps and use this fact to construct a relative algorithm.

Our aim here will be to show how to solve the word problem in free fields, and of the two alternatives described above we shall take the second, thus our solution will not depend on the precise algorithm in K but merely that it exists. In fact it is not enough to assume that K has a solvable word problem; we need to assume that K is dependable over its centre: Given a field K which is a k-algebra, we shall call K *dependable* over k if there is an algorithm which for each finite family of expressions for elements of K, in a finite number of steps leads either to a linear dependence relation between the given elements over k or shows them to be linearly independent over k.

When K is dependable over k, K and hence k has a solvable word problem, as we see by testing 1-element sets for linear dependence. Let K have centre C; our task will be to solve the word problem for the free field $K_C\langle X\rangle$. For this it will be necessary to assume K dependable over C; this assumption is indispensable for we shall see that it holds whenever $K_C\langle X\rangle$ has a solvable word problem.

There is another difficulty which needs to be briefly discussed. As observed earlier we need to deal with *expressions* of elements in a skew field and our problem will be to decide when such an expression represents the zero element. But in forming these expressions we may need to invert non-zero elements, therefore we need to solve the word problem already in order to *form* meaningful expressions.

This problem could be overcome by allowing formal expressions such as $(a - a)^{-1}$; but we shall be able to bypass it altogether: instead of building up rational functions step by step, we can obtain them in a single step by solving suitable matrix equations, as explained in 4.2. In fact we have the following reduction theorem.

Theorem 6.4.1. *Let* R *be a semifir and* U *its universal field of fractions, then the word problem for* U *can be solved if the set of full matrices over* R *is recursive.*

Proof. Any element u_1 of U is obtained as the first component of the solution of a matrix equation

$$Au + a = 0,$$

and $u_1 = 0$ if and only if $A_1 = (a, a_2, \ldots, a_n)$ is non-full. By hypothesis there is an algorithm to decide whether A_1 is full or not, and this provides the answer to our question. ■

We note that it is enough to assume that the set of full matrices over R is recursively enumerable, because its complement, the set of all non-full matrices is always recursively enumerable (in an enumerable ring).

In the same way one can show that for an epic R-field K the word problem can be solved if the set of all matrices inverted over K is recursive.

We now come to the main result to be proved in this section:

Theorem 6.4.2 *Let* K *be a field, dependable over its centre* C, *then the free* K*-field on a set* X *over* C, $U = K_C \langle X \rangle$ *has a solvable word problem and is again dependable over* C. *Conversely, if the word problem in* U *is solvable (for an infinite set* X*), then* K *is dependable over* C.

To prove the theorem we shall at first assume that C is infinite and the degree $[K:C]$ is infinite. Of course this must be understood in a constructive sense: Given $n > 0$, we can in a finite number of steps find n distinct elements of C and n elements of K linearly independent over C. Likewise, any other results we use will need to be put in a constructive form, e.g. the specialization lemma, and Amitsur's theorem on which it was based. Thus given $f \in K_C\langle X\rangle$, $f \neq 0$, there is a method (an 'oracle') for obtaining a set of arguments for which f is non-zero in a finite number of steps.

To prove Th.6.4.2, we must describe an algorithm which will enable us to decide when a matrix A over $F = K_C\langle X\rangle$ is full. First we observe that being full is unaffected by elementary transformations and by taking the diagonal sum with a unit matrix. This allows us to reduce A to a matrix linear in the $x_i \in X$, the process of "linearization by enlargement" (sometimes called 'Higman's trick', cf. Higman [40]). To describe a typical case of this process, suppose that the (n,n)-entry of an n x n matrix has the form f + ab. On enlarging the matrix we can replace the term ab by separate terms a,b by applying elementary transformations, as follows:

$$f + ab \longrightarrow \begin{pmatrix} f + ab & 0 \\ 0 & 1 \end{pmatrix} \longrightarrow \begin{pmatrix} f + ab & a \\ 0 & 1 \end{pmatrix} \longrightarrow \begin{pmatrix} f & a \\ -b & 1 \end{pmatrix}$$

Here only the last two entries in the last two rows are shown. By repeated application we can therefore reduce A to the form

$$(1) \qquad A' = A_o + A_1$$

where A_o is homogeneous of degree 0 and A_1 of degree 1 in the x's. Thus A_o has entries in K and $A_1 = \Sigma B_i x_i$ where the B_i have entries in K; moreover A' is full if and only if A is. Suppose that A' is not full, then it will remain non-full when the x_i are replaced by 0, i.e. A_o must then be singular over K. Thus if A_o is non-singular, A' (and with it A) is necessarily full.

We may therefore suppose that A_o is singular, of rank $r < N$ say, where N is the order of A'. By diagonal reduction over K (which leaves the fullness of A' unaffected) we can reduce A_o to the form $\begin{pmatrix} I & 0 \\ 0 & 0 \end{pmatrix}$ (cf. e.g. Cohn [71"], Ch.8; clearly this is an effective process because K is dependable over C). Let us partition A_1 accordingly, then

$$A' = \begin{pmatrix} I - P & Q \\ R & S \end{pmatrix}$$

where P,Q,R,S are homogeneous of degree 1 (and the sign of P is chosen for convenience in what follows). Now pass to the completion $\hat{F} = K_C\langle\langle X\rangle\rangle$; by the corollary to the inertia theorem A' is full over F if and only if it is full over $\hat{F}$. The matrix $I - P$ is invertible over $\hat{F}$ and by elementary transformations we obtain

$$(2) \quad \begin{pmatrix} I - P & Q \\ R & S \end{pmatrix} \longrightarrow \begin{pmatrix} I & (I - P)^{-1}Q \\ R & S \end{pmatrix} \longrightarrow \begin{pmatrix} I & (I - P)^{-1}Q \\ 0 & S - R(I - P)^{-1}Q \end{pmatrix}.$$

To find whether

$$(3) \quad S - R(I - P)^{-1}Q = 0$$

we have to check that for each $\nu = 0,1,\ldots$ the homogeneous terms of degree ν are 0. Now $S - R(I - P)^{-1}Q = S - \Sigma RP^nQ$ and equating terms of a given degree ν we find that (3) is equivalent to

$$(4) \quad S = 0, \qquad RP^{\nu}Q = 0 \quad (\nu = 0,1,\ldots).$$

These are equations of matrices over F and since the latter is embeddable in a field, we may regard (4) as equations over a field. In that case the equations (4) follow from the same equations with $\nu < N$. Assuming this for now, we thus have an algorithm for determining whether (3) holds. When this equation holds, the matrix on the right of (2) has at least one row of zeros and hence A' is then non-full. If (3) does not hold, then by Lemma 6.3.1 we can specialize the x's within K to values α_i such that $I - P$ remains non-singular and $S - R(I - P)^{-1}Q$ remains non-zero. Translating back to A' we find that specializing to α_i we obtain a matrix of rank $> r$. We now replace x_i by $x_i + \alpha_i$ and start again from (1). This time we have a matrix A_o over K of rank greater than r. By repeating this process a finite number of times (at most N times, where N is the order of A'), we can thus decide whether or not A' is full and this completes the proof in the case where C and $[K:C]$ are infinite.

We still need to prove that the equations (4) all follow from a finite subset; this is related to the well known fact that a nilpotent $n \times n$ matrix A over a field satisfies $A^n = 0$.

Lemma 6.4.3. *Let* P *be an* $n \times n$ *matrix,* Q *a matrix with* n *rows and* R *a matrix with* n *columns over a skew field* K. *If*

$$(5) \qquad RP^{\nu}Q = 0 \text{ for } \nu = 0,1,\ldots,n - 1,$$

then $RP^{\nu}Q = 0$ *for all* ν.

Proof. Let ${}^nK$ be the right K-space of columns with n components. The columns of Q span a subspace V_o of ${}^nK$ while the columns annihilated by the rows of R form a subspace W of ${}^nK$, and since $RQ = 0$ by hypothesis, we have $V_o \subseteq W$. Regarding P as an endomorphism of ${}^nK$ we may define a subspace V_ν of ${}^nK$ for $\nu > 0$ inductively by the equations

$$V_\nu = V_{\nu - 1} + PV_{\nu - 1}.$$

Thus $V_\nu = V_o + PV_o + \ldots + P^\nu V_o$ and it follows that

$$(6) \quad V_o \subseteq V_1 \subseteq \ldots \subseteq V_{n-1}.$$

Moreover, by (5) $V_\nu \subseteq W$ for $\nu = 0,1,\ldots,n-1$. Now if $Q = 0$ or $R = 0$ there is nothing to prove. Otherwise $V_o \neq 0$, $W \neq V$ and we must have equality at some point in (6), since dim $V_{n-1} \leq n - 1$. Suppose that $V_{k-1} = V_k$ $(k < n)$, then $PV_{k-1} \subseteq V_{k-1}$, hence $PV_k \subseteq PV_{k-1} \subseteq V_k$, therefore $V_{k+1} = V_k + PV_k = V_k$ so the sequence is stationary from V_{k-1} onwards. Since the sequence has become stationary by the (n-1)th stage at the latest, we conclude that $V_\nu \subseteq W$ for all ν, i.e. $RP^\nu Q = 0$ for all ν, as we wished to show. ■

We now show how to modify the proof of Th.6.4.2 when C is finite, or even when it is not 'constructively' infinite in the way described earlier. Given a field K with centre C, we form the polynomial ring $K[t]$ in a central indeterminate t and let $K' = K(t)$ be its field of fractions. By Lemma 6.3.5 the centre of K' is $C' = C(t)$. We claim that K' is dependable over C' whenever K is dependable over C. For let $u_1,\ldots,u_n \in K'$ and write these elements as rational functions in t with a common denominator: $u_i = f_i g^{-1}$, where $f_i, g \in K[t]$. Clearly it will be enough to test $f_1,\ldots,f_n$ for linear dependence over $C' = C(t)$. We may take the f's to be numbered so that $\deg f_1 \geq \deg f_2 \geq \ldots$. Consider the leading coefficients of $f_1,\ldots,f_n$; if they are linearly independent over C then the f's are linearly independent over C'. Otherwise we can find i, $1 \leq i \leq n$, $\alpha_{i+1},\ldots,\alpha_n \in C$ and positive integers $\nu_{i+1},\ldots,\nu_n$ such that $f_i' = f_i - \Sigma_{i+1}^n f_j \alpha_j t^{\nu_j}$ has lower degree than f_i. Now the linear dependence over C' of $f_1,\ldots,f_n$ is equivalent to that of $f_1,\ldots,f_{i-1},f_i',f_{i+1},\ldots,f_n$ and here the sum of the degrees is smaller. Using induction on the sum of the degrees of the f's we obtain the result.

Now return to Th.6.4.2. If in that theorem C is finite or more generally, there is no constructive process of obtaining infinitely many elements in C, we adjoin a central indeterminate t so that K, C are replaced by $K' = K(t)$, $C' = C(t)$. By what we have just seen, K' is dependable over C' whenever K is so over C. Moreover C' is infinite in the constructive sense (e.g. we can take $1, t, t^2, \ldots$) and $[K':C'] = [K:C]$. It follows that the set of full matrices over $K'_{C'}\langle X\rangle$ is recursive, hence by Th.6.3.6 (or even the special case Lemma 6.3.4) the set of all full matrices over $K_C\langle X\rangle$ is also recursive. Hence the word problem in U is soluble, so Th. 6.4.2 continues to hold even when the centre of K is finite.

There remains the case where K has finite degree over its centre, or where no process for constructing infinite linearly independent sets exists. Instead of adjoining a central indeterminate we now form a skew extension.

Let K be a field with centre C and let σ be an automorphism of K leaving C elementwise fixed. We form the skew polynomial ring $R = K[y;\sigma]$ and its field of fractions $K' = K(y;\sigma)$. If no power of σ is an inner automorphism, then the centre of K' is C. To see this we embed K' in the field of skew Laurent series $K((y;\sigma))$ (cf.2.1). If $f = \Sigma y^\nu a_\nu$ lies in the centre then $fy = yf$, hence $a_\nu^\sigma = a_\nu$ and $cf = fc$ for all $c \in K$, hence $c^{\sigma^\nu} a_\nu = a_\nu c$. Since σ^ν is not inner for $\nu \neq 0$, it follows that $a_\nu = 0$ except when $\nu = 0$ and $a_0 c = c a_0$, i.e. $f = a_0 \in C$. Clearly K' is of infinite degree over its centre, e.g. the powers of y are linearly independent.

Taking $C' = C$ in Th.6.3.6 we find that the embedding

$$K_C\langle X\rangle \longrightarrow K'_C\langle X\rangle \tag{7}$$

is honest. Moreover if K is dependable over C (and if σ is 'computable' in an obvious sense) then so is K'. For let $u_1,\dots u_n \in K'$; as before we write $u_i = f_i g^{-1}$, where $f_i, g \in K[y;\sigma]$ and it is again enough to test $f_1,\dots,f_n$ for linear dependence over C. This time we single out the f's of maximal degree, $f_1,\dots,f_r$ say. If their leading terms are linearly independent over C, then so are the f's. Otherwise let $f_i' = f_i - \Sigma_{i+1}^r f_j\alpha_j$ $(\alpha_j \in C)$ have lower degree than f_i and continue with $f_1,\dots,f_i',\dots,f_n$ as before; again this process ends after a finite number of steps.

To complete the proof of Th.6.4.2 we shall need another Lemma which will also establish the second part of the theorem.

Lemma 6.4.4. *Let* K *be a field (over* k*). If for every finite set* Y *the word problem for the free* K*-field on* Y *over* k *is soluble, then the free* K*-field on any set* X *over* k *is dependable over* k.

Proof. Let $U = K_k \langle X \rangle$ be the free field; we may assume X infinite by embedding U in a free K-field on an infinite set containing X (that such an embedding exists follows from Th.6.3.6 but is also easy to see directly). Given $u_1,\dots,u_n \in U$, we have to determine whether the u's are linearly independent over k. We shall use induction on n, the case n = 1 being essentially the word problem for U. We may assume $u_1 \neq 0$, and hence on dividing by u_1 we may suppose that $u_1 = 1$. Only finitely many elements of X occur in $u_2,\dots,u_n$, so we can find another element in X, y say. Write $u_i' = u_i y - y u_i$ and check whether $u_2',\dots,u_n'$ are linearly dependent over k. If so, let $\Sigma_2^n u_i'\alpha_i = 0$, where $\alpha_2,\dots,\alpha_n \in k$ and are not all zero, then $u = \Sigma_2^n u_i\alpha_i$ satisfies $yu = uy$. Since u does not involve y, it follows that u represents an element α of k (which can be computed by suitably specializing the x's), and hence $1.\alpha - \Sigma_2^n u_i\alpha_i = 0$ is a dependence relation over k. Conversely, if there is a dependence relation $\Sigma_1^n u_i\alpha_i = 0$, where not all the α_i vanish, then not all of $\alpha_2,\dots,\alpha_n$ can vanish (because $u_1 =$

$1 \neq 0$), and so $\Sigma_2^n u_i' \alpha_i = 0$ is a dependence relation between $u_2', \ldots, u_n'$. The result now follows by induction on n. ■

We note that since K is a subfield of U, K is dependable over C; thus the dependability of K is a consequence of the solubility of the word problem for U (on an infinite set). This completes the proof of Th.6.4.2 when $[K:C]$ is infinite.

When $[K:C]$ is finite, but K has a (computable) automorphism over C, no power of which is inner, we can form the skew function field $K' = K(y;\sigma)$. Then K' is of infinite degree over its centre C and K' is dependable over C, hence the set of all full matrices over $K'_C\langle X\rangle$ is recursive, and so is the set of all full matrices over $K_C\langle X\rangle$, because (7) is honest. This solves the word problem for U.

Finally if K has no automorphism σ of the required kind, we form $K' = K(t)$ with a central indeterminate t; as we have seen, the result is a field K' with centre C(t). Now repeat the process with the endomorphism $f(t) \longmapsto f(t^2)$. This is computable (in any reasonable sense) and it induces an automorphism of $K(t, t^{\frac{1}{2}}, t^{\frac{1}{4}}, \ldots)$, no power of which is inner, hence we obtain a field K'' of infinite degree over its centre C.

From Th.6.3.6 we obtain the following special case by taking $K = C$:

Corollary. Let k *be any commutative field with soluble word problem, then* $U = k\langle X\rangle$ *has soluble word problem.* ■

There still remains the word problem for a free field $K_k\langle X\rangle$ where k is not the exact centre of K. This really requires a sharper form of the specialization lemma, but we shall not pursue the matter here.

6.5 A skew field with unsolvable word problem

As is to be expected, for general skew fields the word problem is unsolvable; an example was given by Macintyre [73]. The account below is a (slightly simplified) version of another example due to Macintyre. The idea is to take a finitely presented group with unsolvable word problem and use these relations in the group algebra of the free group. We need a couple of preparatory lemmas.

Lemma 6.5.1. *Let* F_x *be the free group on* $x_1,\ldots,x_n$ *and* F_y *the free group on* $y_1,\ldots,y_n$. *In the direct product* $F_x \times F_y$ *let* H *be the subgroup generated by the elements* $x_i y_i$ $(i = 1, \ldots, n)$ *and elements* $u_1,\ldots,u_m \in F_x$. *Then* $H \cap F_x$ *is the normal subgroup of* F_x *generated by* $u_1,\ldots,u_m$.

Proof. Let N be the normal subgroup of F_x generated by the u_μ $(\mu = 1,\ldots,m)$. Since $x_i^{-1} u_\mu x_i = (x_i y_i)^{-1} u_\mu (x_i y_i) \in H$, it is clear that $N \subseteq H$, hence $N \subseteq H \cap F_x$. To prove equality, consider the obvious homomorphism

$$f : F_x \times F_y \longrightarrow (F_x/N) \times F_y$$

which maps u_μ to 1 $(\mu = 1,\ldots,m)$. If $w \in H \cap F_x$, then wf is a product of the $(x_i y_i)f$, and since the x's and y's commute, we can write it as

$$wf = [v(x)v(y)]f = v(xf)v(yf),$$

where v is a word in n symbols. Since $w \in F_x$, $wf \in F_x/N$ and so $v(yf) = 1$, but the $y_i f$ are free, so v is the empty word and $wf = 1$, hence $w \in \ker f = N$. ∎

Let F_x, F_y, H, N be as above and consider $G = F_x \times F_y$. This group can be ordered: we order the factors as in 2.1 and then take the lexicographic order on G. Hence we can form the power series field $K = k((G))$. The power series with support in H form a subfield L; we take a family of copies of K indexed by **Z** and form their coproduct amalgamating L. The resulting ring is a fir, with universal field of fractions D say. If σ is the shift automorphism, we can form the field of fractions $D(t;\sigma)$ of the skew polynomial ring $D[t;\sigma]$.

Lemma 6.5.2. *With the above notation, let* $w \in F_x$, *where* $F_x \subseteq G \subseteq K = K_o \subseteq D$; *then* $w \in N$ *if and only if* $wt = tw$ *in* $D(t;\sigma)$.

Proof. If $w \in N$, then $w \in H$ by Lemma 6.5.1, hence $w \in L$ and so $wt = tw$.

Conversely, if $tw = wt$, w is fixed under σ and so lies in the fixed field of σ, i.e. $w \in L$. But L consists of all power series with support in H, so $w \in H \cap F_x = N$, by Lemma 6.5.1. ∎

Now let A be a finitely presented group with unsolvable word problem, say

$$(1) \qquad A = gp\{x_1,\ldots,x_n \mid u_1 = \ldots = u_m = 1\},$$

where $u_1,\ldots,u_m$ are words in the x's. We shall construct a finitely presented field whose word problem incorporates that of A. Let

$$M = k \langle x_1,\ldots,x_n,y_1,\ldots,y_n,t \mid \Phi \rangle,$$

where Φ consists of the following equations:

$$(2) \qquad x_iy_j = y_jx_i \ (i,j = 1,\ldots,n), \quad x_iy_it = tx_iy_i \ (i = 1,\ldots,$$

$$u_\mu t = tu_\mu \ (\mu = 1,\ldots,m).$$

To see that this is meaningful, let $P = P_x$ be the free field over k on $x_1,\ldots,x_n$ and form

$$P\langle y_1,\ldots,y_n\rangle = P[y_1] \sqcup_P \cdots \sqcup_P P[y_n].$$

This is a fir and so has a universal field of fractions Q; moreover, $F_x \times F_y$ is naturally embedded in Q, in fact Q is also the universal field of fractions of the group algebra of $F_x \times F_y$ over k. In Q consider the subfield R generated over k by H, and let S be the field coproduct of copies of Q indexed by **Z** amalgamating R. If σ is the shift automorphism in S, we can form $T = S(t;\sigma)$; from its construction this is essentially M (cf. Lemma 5.5.4). By the universality of T we have a specialization from T to $D(t;\sigma)$. We claim that

$$(3) \qquad w \in N \iff tw = wt \text{ in } T \quad \text{for any } w \in F_x.$$

$\Rightarrow$ is clear, to prove $\Leftarrow$, we note that if $tw = wt$ in T, then this also holds in $D(t;\sigma)$, hence $w \in N$ by Lemma 6.5.2.

Now (3) shows that the word problem in M(= T) is unsolvable because this is the case for $F_x/N = A$.

7 · Rational relations and rational identities

7.1 Polynomial identities

Every ring satisfies certain identities such as the associative law: $(xy)z = x(yz)$. In a field the situation is less simple; we have rational identities like $xx^{-1} = 1$ or $(xy)^{-1} = y^{-1}x^{-1}$, but here it is necessary to restrict x,y to be different from 0. In order to discuss rational identities over a field it is helpful to summarize the situation for rings.

Let k be a commutative field and $F = k\langle X\rangle$ the free k-algebra on a set $X = \{x_1, x_2, \ldots\}$. Any k-algebra A is said to satisfy the polynomial identity

$$(1) \quad p(X) = 0,$$

if p is an element of F which vanishes for all values of the set X in A. If A satisfies a non-trivial identity (i.e. if $p \neq 0$) it is called a *PI-algebra*. The basic result on PI-algebras is

Kaplansky's theorem. Let R *be a primitive* PI-*algebra with an identity of degree* d. *Then* R *is a simple algebra of finite dimension* n^2 *over its centre, where* $n \leq d/2$.

For a proof see Jacobson [56], Herstein [68] or Cohn [77].

In (1) the coefficients were restricted to lie in the centre; without this restriction the result clearly fails to hold. E.g. in any matrix ring $\mathfrak{M}_n(k)$ over a commutative field we have

$$e_{11}xe_{11}ye_{11} - e_{11}ye_{11}xe_{11} = 0.$$

This has degree 2, but it is not an identity of the form (1), and the dimension of $\mathfrak{M}_n(k)$ over k is n^2. Nevertheless the existence of such a generalized identity limits the algebra severely, as Amitsur has shown (cf. 6.3 above). For reference we state

Amitsur's theorem. Let R *be a primitive ring, then* R *satisfies a generalized polynomial identity (cf.* 6.3*) if and only if it is isomorphic to a dense ring of linear transformations over a skew field* K *finite-dimensional over its centre, and* R *contains a linear transformation of finite rank.*

The connexion between generalized and ordinary polynomial identities has been described by Procesi [68]; let us recall the details.

Let D be a simple algebra of finite dimension n over its centre k, then k is a field, n is a perfect square, say $[D:k] = n = d^2$ and $[D^m:k] = mn$, for $m \geq 1$. In terms of a k-basis $u_1, \ldots, u_n$ for D we can write the elements of the free D-ring $F = D_k\langle x_1, \ldots, x_m\rangle$ as linear combinations (over k) of monomials $u_{i_0} x_{j_1} u_{i_1} \ldots x_{j_r} u_{i_r}$. There is a pairing

$$F \times D^m \longrightarrow D,$$

defined by $(f,a) \longmapsto f(a)$, where $f(a) = f(a_1, \ldots, a_m)$ arises from $f(x_1, \ldots, x_m)$ on replacing x_i by $a_i \in D$. If we fix $f \in F$, we have a mapping $D^m \longrightarrow D$ and if we fix $a \in D^m$ we get a mapping $F \longrightarrow D$. Clearly the latter mapping is a D-ring homomorphism. Let us write Hom(F,D) for the set of all D-ring homomorphisms, then we have

Lemma 7.1.1. *If* D *is a* k*-algebra and* $F = D_k\langle x_1, \ldots, x_m\rangle$, *then*

$$\mathrm{Hom}(F,D) \cong D^m.$$

Explicitly we have $\phi \longmapsto (x_1^\phi, \ldots, x_m^\phi)$.

This follows immediately: we have seen that $a \in D^m$ defines a homomorphism and conversely, each homomorphism ϕ provides an element $(x_1^{\phi},\ldots,x_m^{\phi}) \in D^m$. ■

This just expresses the left adjoint property of F, and here the precise nature of D is quite immaterial.

In a similar way we get a mapping Θ, the evaluation mapping

$$(2) \qquad \Theta : F \longrightarrow D^{D^m}, \qquad f \longmapsto \bar{f},$$

where $\bar{f}: D^m \longrightarrow D$ is the function on D^m defined by f. Now (2) is also a D-ring homomorphism if we regard the functions from D^m to D as a ring under pointwise operations; this amounts to treating the right-hand side of (2) as a product of rings. The image of F under Θ is written $\bar{F}$; it is the ring of polynomial functions in m variables on D. The kernel of Θ is just the set of all generalized polynomial identities in m variables on D.

By identifying D^m with k^{nm} via a k-basis of D, we may view the ring of polynomial functions $k^{nm}(= D^m) \longrightarrow k$ as a central k-subalgebra G of D^{D^m}; clearly G does not depend on the choice of k-basis of D. Since the canonical map $k^{D^m} \otimes_k D \longrightarrow D^{D^m}$ is injective, the subring C of D^{D^m} generated by D and G is of the form $G \otimes_k D$. If moreover, k is infinite, G is just the k-algebra of polynomials in mn commuting indeterminates, so C is the D-ring of polynomials in mn central indeterminates. Another description of C is given in

Theorem 7.1.2. *Let* D *be an* n-*dimensional central simple* k-*algebra, then* $F = D_k\langle x_1,\ldots,x_m\rangle$ *may be expressed as the free* D-*ring on* mn D-*centralizing indeterminates and* $C = \bar{F}$ *is the image of the evaluation map* Θ.

Proof. We may regard F as the tensor D-ring on the D-bimodule $(D \otimes_k D)^m$, and since D is central simple, the map

(4) $\phi : D \otimes_k D \longrightarrow \mathrm{End}_k(D) \cong D^n$, where $(a \otimes b)\phi : x \longmapsto axb$

is a D-bimodule isomorphism. It follows that F is the tensor D-ring on mn D-centralizing indeterminates.

Now fix a k-basis $u_1,\ldots,u_n$ of D and consider the dual k-basis $u_1^*,\ldots,u_n^* \in \mathrm{Hom}_k(D,k) \subseteq \mathrm{End}_k(D)$. For each $\mu = 1,\ldots,n$ there exists $v_\mu = \Sigma a_{\mu\lambda} \otimes u_\lambda \in D \otimes D$ mapping onto u_μ^*, i.e. such that $v_\mu\phi = u_\mu^*$. If in $F = D_k\langle x_1,\ldots,x_m\rangle$ we take $v_{i\mu} = \Sigma a_{\mu\lambda}x_i u_\lambda$, then $F = D\langle v_{i\mu}\rangle$ is the free D-ring on the mn D-centralizing indeterminates $v_{i\mu}$, $i = 1,\ldots,m$, $\mu = 1,\ldots,n$. Write $\xi_{i\mu} = v_{i\mu}\Theta$: $(\Sigma_\lambda b_{1\lambda}u_\lambda,\ldots,\Sigma_\lambda b_{m\lambda}u_\lambda) \longmapsto b_{i\mu}$, then it is clear that $\bar{F}$ is the D-ring generated by the $\xi_{i\mu}$. Now G is by definition the k-algebra generated by the $\xi_{i\mu}$, hence $C = GD = \bar{F}$ as claimed. ∎

If we examine the role played by Θ we obtain

Theorem 7.1.3. *If* k *is infinite and* D *is an* n*-dimensional central simple* k*-algebra, then the evaluation map* Θ *can be expressed in the form*

$$F = D\langle v_{i\mu}\rangle \longrightarrow D[\xi_{i\mu}] = C \subseteq D^{D^m}, \text{ where } v_{i\mu} \longmapsto \xi_{i\mu}$$

$$(i = 1,\ldots,m; \mu = 1,\ldots,n)$$

hence the kernel of Θ *is generated by the commutators of pairs of the* $v_{i\mu}$. ∎

The above account follows Procesi [68], with simplifications by Dicks; cf. also Gordon-Motzkin [65], who prove Th.7.1.2 when D is a field. For a more general treatment, in the context of Azumaya algebras, see Procesi [73].

7.2 Rational identities

The basic result on rational identities, again due to Amitsur [66], states that there are no non-trivial rational identities over a skew field which is infinite-dimensional over its centre and which has an infinite centre. But it is now more tricky to decide what constitutes a 'non-trivial' identity. Here are some 'trivial' ones:

$$(x + y)^{-1} = y^{-1}(x^{-1} + y^{-1})^{-1}x^{-1},$$

$$\left[x^{-1} + (y^{-1} - x)^{-1}\right]^{-1} = x - xyx \qquad \text{(Hua's identity).}$$

We shall give two proofs of Amitsur's result, one by Bergman [70] and one by the author, based on the results of 6.3 above (cf. Cohn [72']).

Our first task is to find a means of expressing rational functions; here we shall follow Bergman [70,76]. Let D be a skew field with centre k, then we can form D(t), the field of rational functions in a central indeterminate. Any $\phi \in D(t)$ has the form $\phi = fg^{-1}$, where f,g are polynomials in t, and we can set $t = \alpha \in k$ if α is such that $g(\alpha) \neq 0$; then $\phi(\alpha)$ will be defined. Given ϕ, we can choose f,g in $\phi = fg^{-1}$ to be coprime, and then f,g will not both vanish for any $\alpha \in k$. Since we only had to avoid the zeros of g in defining $\phi(\alpha)$ we see that ϕ is defined at all but finitely many points of k.

We have to generalize this to the case of several non-central variables. Now we no longer have D(t) at our disposal, and although we have seen in Ch.6 how to construct free fields, that construction will not be needed here. What we shall do is to build up formal expressions in $x_1, \ldots, x_m$ using $+, -, \times$, $^{-1}$ and elements of D. The expressions will be defined on a subset of D^m or more generally on E^m, where E is a D-field.

Let X be any set. An X-*ring* is a ring R with a mapping $\alpha: X \longrightarrow R$; we write R or (R,α) to emphasize the mapping. If R is a field we speak of an X-*field*; this is essentially the same as a $\mathbf{Z}\langle X\rangle$-field in our previous terminology.

Given $X = \{x_1,\ldots,x_m\}$ we write $\mathbf{R}(X)$ for the free abstract algebra on X with operations $\{0_o, 1_o, -_1, ()^{-1}_1, +_2, \times_2\}$, where subscripts indicate the arity of the operation. For each expression there is a unique way of building it up since no relations are imposed, thus e.g. $(x-x)^{-1}$ exists. In contrast to 7.1 we now have a *partial* evaluation mapping

$$(1) \qquad \mathbf{R}(X) \times R^m \longrightarrow R.$$

Thus any map $\alpha : X \longrightarrow R$ defines a map $\bar{\alpha}$ of a subset of $\mathbf{R}(X)$ into R, by the following rules:

(i) if $a = 0$ or 1, $a\bar{\alpha} = 0$ or 1,

(ii) if $a = x_i$, $x_i\bar{\alpha} = x_i\alpha$,

(iii) if $a = -b$ or $b+c$ or bc and $b\bar{\alpha}$, $c\bar{\alpha}$ are defined, then $a\bar{\alpha} = -b\bar{\alpha}$ or $b\bar{\alpha} + c\bar{\alpha}$ or $b\bar{\alpha}.c\bar{\alpha}$.

(iv) if $a = b^{-1}$ and $b\bar{\alpha}$ is defined and invertible in R, then $a\bar{\alpha} = (b\bar{\alpha})^{-1}$.

Since $\bar{\alpha}$ just extends α we can safely omit the bar. In a field invertible is the same as non-zero, hence we have

Proposition 7.2.1. *Let* X *be a set,* (D,α) *an* X*-field and* $a \in \mathbf{R}(X)$*, then* $a\alpha$ *is undefined if and only if* a *has a subexpression* b^{-1}*, where* $b\alpha = 0$. ■

With each $\alpha : X \longrightarrow D$ we can associate a subset E(D) of $\mathbf{R}(X)$, the *domain* of α, consisting of expressions which can be evaluated for α. Similarly with each $f \in \mathbf{R}(X)$ we associate its domain dom f, a subset of D^m consisting of the points at which f is defined; more generally we shall consider dom f in E^m, where E is a D-field. If dom f $\neq \emptyset$, f is called *non-degenerate* on E. In this section we shall mainly be dealing with the domains of functions $f \in \mathbf{R}(X)$.

Lemma 7.2.2. *Let* D *be a skew field which is an algebra*

over an infinite field k. *If* f,g *are non-degenerate on a* D-*field* E *then* dom f $\cap$ dom g $\neq \emptyset$.

Proof. Let $p \in$ dom f, $q \in$ dom g, write $r = tp + (1-t)q$ and consider $f(r)$, $g(r) \in E(t)$. Each is defined for all but finitely many values of t in k, hence for some $\alpha \in k$ both are defined. ■

Given $f,g \in \mathbf{R}(X)$, let us put $f \sim g$ if f,g are non-degenerate (on a given E) and f,g have the same value at each point of dom f $\cap$ dom g. This is clearly an equivalence, the transitivity follows by Lemma 7.2.2. If f,g are non-degenerate, so are f+g, f-g, fg; moreover they depend only on the classes of f,g not on f,g themselves, and if $f \not\sim 0$, then f^{-1} is defined. Thus we have

Theorem 7.2.3. *Let* D *be a skew field with infinite centre* k *and* E *a* D-*field which is also a* k-*algebra, then the equivalence classes of rational functions from* E^m *to* E *with coefficients in* D *form a skew field* $D_E(X)$. ■

If E is commutative, this reduces to D(X) and is independent of E. In that case any element of D(X) can be written as a quotient of two coprime polynomials, and this expression is essentially unique. The dependence on E in the general case will be examined below; now there is no such convenient normal form for the elements of $D_E(X)$. Even if we use Ch.6, a given element may satisfy more than one matrix equation Au = a and the relation between them remains to be described (cf. Cohn [b]). In terms of explicit rational expressions for f, it may be that different expressions have different domains and the resulting function is defined on the union of these domains. Bergman [70] raises the question: "Whether there is always an expression for f having this whole set for domain of definition, a 'universal' expression for the rational function f."

Generally the domains of functions form a basis for the open sets of a topology on E^m, the *rational topology* on E^m

(cf. also 8.5 below; the polynomial topology, a priori coarser, is the Zariski topology). The closed sets are of the form

$$V(P) = \{p \in E^m \mid f(p) = 0 \text{ for all } f \in P\}$$

where $P \subseteq D_E(X)$.

A subset S of E^m is called *irreducible* if it is non-empty and not the union of two closed proper subsets. Equivalently: the intersection of non-empty open subsets of S is non-empty. Thus Lemma 7.2.2 states that E^m is irreducible in the rational topology when the centre of E is infinite.

A subset S of E^m is called *flat* if $p,q \in S$ implies $\alpha p + (1-\alpha)q \in S$ for infinitely many $\alpha \in k$. Of course a closed flat subset will then contain $\alpha p + (1 - \alpha)q$ for all $\alpha \in k$. Now the proof of Lemma 7.2.2 gives us

Lemma 7.2.4. *Any non-empty flat subset of* E^m *is irreducible.* ■

An example of a flat closed subset is the space S defined by

$$\Sigma a_{i\lambda} x_i b_{i\lambda} = c \qquad (a_{i\lambda}, b_{i\lambda}, c \in D). \tag{2}$$

By Lemma 7.2.4, S is irreducible (if non-empty) and so as in Th. 7.2.3 yields a skew field $D_S(X)$ in $x_1,\ldots,x_m$ satisfying (2), the function field of (2). In general it is not easy to decide whether a given set is irreducible, e.g. $x_1x_2 - x_2x_1 = 0$ for $E \supseteq D \supseteq k$. In the commutative case every closed set is a *finite* union of irreducible closed sets, but this need not hold in general.

It is clear that polynomially closed $\Rightarrow$ rationally closed; we want conditions for the converse to hold. First two remarks:

(i) Let $S \subseteq E^m$ be such that $p \notin \bar{S}$, then there exists f defined at p but not anywhere on S. The degeneracy of f can only arise by inversion, so $f = g^{-1}$, where g is non-degenerate on S and 0 at all points of S where defined, and $g(p) \neq 0$.

(ii) Any element of D(t) defined at t = 0 can be expanded in a power series. Let g = a - th say, then $g^{-1} = a^{-1}(1 - tha^{-1})^{-1} = \Sigma a^{-1}(tha^{-1})^n$. So we can build up any function in D(t) provided that it is defined at t = 0.

Lemma 7.2.5. *Let* D *be a skew field which is a* k-*algebra, where* k *is an infinite field, and let* E *be a* D-*field and a* k-*algebra. If* $S \subseteq D^m$ *is flat, then its closure in* E^m *is polynomially closed.*

Thus for flat sets, rationally closed = polynomially closed.

Proof. Let $p \notin \bar{S}$; we have to find a polynomial over D which is zero on S but not at p. We know that there is a rational function f, non-degenerate on S and f = 0 on S but $f(p) \neq 0$. Say f is defined at $q \in S$.

For any $x \in E^m$ consider $f((1-t)q + tx)$; this is defined for t = 0, so it is a well-defined element of E(t). If $x \in S$, f is 0 by flatness, but for x = p it is non-zero because it is non-zero for t = 1. In the power series expansion of $f((1-t)q + tx)$, if we have to take the inverse of an expression h(t), the constant term h(0) is non-zero, because f(q) is defined, and h(0) does not involve the coordinates of x. So the expansion $f((1-t)q + tx)$ has coefficients which are polynomials in x; their coefficients are in D because $q \in S \subseteq D^m$. These polynomials are 0 on S, but at least one is non-zero at p, and this is the required polynomial.■

Corollary. Let k,D,E *be as before, and assume that* D *and* E *satisfy the same generalized polynomial identities over* **k** *with coefficients in* D, *then* $D_E(X) \cong D_D(X)$ *and for any* *k-subfield* C *of* D, $C_E(X) \cong C_D(X)$.

Proof. The rational closure of D^m in E^m is polynomially closed by Lemma 7.2.5. Now every g.p.i. on D^m holds on E^m, so the rational closure of D^m is E^m, i.e. D^m is dense in E^m. The rest follows because $C_E(X)$ is the subfield of $D_E(X)$ generated by C and X. ∎

Now the rational identities may be described as follows:

Theorem 7.2.6. *(Bergman* [70]*). Let* D *be a field with centre* k *then there exists a* D*-field* E *with infinite centre* $C \supseteq k$ *such that* $[E:C]=\infty$, *and for each* m, *any such* E, C *yield the same function field* $D_E(X)$ $(m = |X|)$.

Proof. The last part follows from Lemma 7.2.5; it only remains to produce E,C. If k is finite, adjoin t to get $D_1 = D(t)$. Now let $F = D_1(u)$ with endomorphism $\alpha:f(u) \longmapsto f(u^2)$ and form $E = F(v;\alpha)$. As in the proof of Th.6.3.6 we see that the centre of E is $k\,(t)$. ∎

This theorem may be expressed by saying that for skew fields infinite over their centre (where the latter is infinite) there are no non-trivial rational identities. We now give another proof of this result using the methods of Ch.4 and Th.6.3.6.

Theorem 7.2.7. *Let* E *be a field with centre* C, *let* D *be a subfield of* E *and write* $k = D \cap C$. *Assume further that* (i) $[E:C] =\infty$ *and* (ii) C *is infinite. Then every element of* $D_k\langle X\rangle$ *is non-degenerate on* E.

Proof. We know that any $p \in D_k\langle X\rangle$ can be obtained as a component $p = u_1$ of the solution of an equation

$$Au = a,$$

where A is a full matrix over $D_k\langle X\rangle$, and p will be non-degenerate on E provided that A goes over into a non-singular matrix under some substitution $X \longrightarrow E$. By Th.6.3.6, the mapping $D_k\langle X\rangle \longrightarrow E_C\langle X\rangle$ is honest, hence A is full over

$E_C\langle X\rangle$ and by the specialization lemma 6.3.1, A can be specialized to a non-singular matrix over E, which is what was needed. ■

When D is finite-dimensional over its centre, there are of course non-trivial identities, but Amitsur [66] shows that they depend only on the degree (cf. also Bergman [70]). More precisely, if $[D:k] = n = d^2$ and E is any extension of D with infinite centre C containing k, where $[E:C] = (rd)^2$, then $D_E(X)$ depends only on D,d,r, m = $|X|$ and not on E. It is shown that $D_E(X)$ has dimension $(rd)^2$ over its centre, hence these fields are different for different values of rd. Moreover for $d_1|d_2$ the field with d_1 is a specialization of that with d_2 (cf. Bergman [70]).

7.3 Specializations

We now examine how rational identities change under specialization. Of course we must first define the appropriate notion of specialization. A homomorphism between two rings is said to be *local* if it maps non-units to non-units. Let D, D' be fields, then a local homomorphism from a subring D_1 of D to D' is also called a *local homomorphism* from D to D' with *domain* D_1. If $\phi:D \longrightarrow D'$ is a local homomorphism with domain D_1, then ker ϕ is the set of non-units of D_1, hence D_1 is a local ring with residue class field $D_1/\ker \phi$ isomorphic to a subfield of D'.

Let (D,α), (D',α') be X-fields, then a local homomorphism $\phi:D \longrightarrow D'$ whose domain contains X^α and such that $\alpha' = \alpha\phi$ is called an X-*specialization*. Clearly this exists only if the domain of α contains that of α'.

To describe rational identities we shall need the notion of PI-degree. Let A be a commutative ring, then $\mathfrak{M}_n(A)$ is n^2-dimensional over its centre (as free A-module) and it satisfies the standard identity of degree 2n (Amitsur-Levitzk theorem):

$$S_{2n}(x_1,\ldots,x_{2n}) = \Sigma \operatorname{sgn} \sigma \; x_{1\sigma}x_{2\sigma}\ldots x_{(2n)\sigma} = 0.$$

Let R be any prime PI-ring; by Posner's theorem it has a ring of fractions Q which is simple Artinian and satisfies the same polynomial identities as R (cf. Jacobson [75] or Cohn [77]). Let Q be d^2-dimensional over its centre, then R satisfies $S_{2d} = 0$ and no standard identity of lower degree. We shall call d the PI-*degree* of R (and Q) and write d = PI-deg R. If R is a prime ring satisfying no polynomial identity its PI-degree is said to be ∞.

We shall also need the notion of generic matrix ring. Let k be a commutative field and $m,d \geq 1$. Write $k[T]$ for the commutative polynomial ring over k in the family $T = \{x_{ij}^{\lambda}\}$ of commuting indeterminates, where $i,j = 1,\ldots,d$, $\lambda = 1,\ldots,m$. Let k(T) be its field of fractions and consider the matrix rings

$$\mathfrak{M}_d(k[T]) \subseteq \mathfrak{M}_d(k(T)).$$

We have a canonical m-tuple of matrices $X_\lambda = (x_{ij}^{\lambda})$; the k-algebra generated by these m matrices is written $k\langle X\rangle_d$ and is called the *generic matrix ring* of order **d**. It is the free k-algebra on $X = \{X_\lambda\}$ in the variety of k-algebras generated by d x d matrix rings over commutative k-algebras. Amitsur has shown that $R = k\langle X\rangle_d$ is entire (cf. e.g. Cohn [77]: one has to find a field of PI-degree d and embed it in $\mathfrak{M}_d(E)$, where $E \supseteq k$). As an entire PI-ring R is an Ore domain; its field of fractions is written $k\langle\!\langle X\rangle\!\rangle_d$; like $k\langle X\rangle_d$ it has PI-degree d, if $m > 1$. Of course for $m = 1$, $k\langle X\rangle_d$ reduces to a polynomial ring in one variable; this is not of interest and we henceforth assume that $m > 1$.

Let (D,α) be any X-field; we defined in 7.2 its domain E(D) as the subset of **R**(X) for which α is defined. Let Z(D) be the subset of E(D) consisting of all functions which vanish for α. Any $f \in Z(D)$ is called a *rational relation*, or k-rational relation if coefficients in k are allowed. Explicitly, we have $f^{\alpha} = 0$ in D, but of course this presupposes that f^{α} is defined. Now Amitsur's theorem on rational identities (7.2) may be expressed as follows: Let D be a field with infinite centre k, then there is an X-field E over k such that the k-rational identities over D are the k-rational relations satisfied by X over E. Thus we can speak of E as the *free* X-field for this set of identities. Moreover, the structure of E depends only on k, m and the PI-degree of D:

If PI-deg D = d, then $E = k \langle X \rangle_d$ is the field of generic matrices,

If PI-deg D $=\infty$, then $E = k\langle X \rangle$ is the free k-field on X.

In particular, two k-fields satisfy the same rational identities if and only if they have the same PI-degree. For our first theorem we need a result of Bergman-Small [75]. We recall that a ring R is *local* if $R/J(R)$ is a field (where $J(R)$ is the Jacobson radical of R); if $R/J(R)$ is a full matrix ring over a field, R is said to be a *matrix local ring*.

Theorem 7.A *(Bergman-Small* [75]*)*

(i) *If* R *is a prime* PI-ring *which is also local (or even matrix local) with maximal ideal* $\mathfrak{m}$ *then* PI-deg $R/\mathfrak{m}$ *divides* PI-deg R.

(ii) *If* $R_1 \subseteq R$ *are* PI-*domains, then* PI-deg R_1 *divides* PI-deg R.

We shall sketch the proof of (ii) only. Let d_1, d be the PI-degrees of R_1, R. They are also the PI-degrees of their fields of fractions Q_1, Q. Let k_1, k be their centres; by enlarging Q_1 we may assume that $k_1 \supseteq k$. Now choose a maximal commutative subfield F_1 of Q_1 and enlarge F_1 to a maximal commutative subfield F of Q, then $[F_1:k_1]$ divides $[F:k]$, and this means that $d_1 | d$. ■

With the help of this result we can describe the specializations between generic matrix rings, following Bergman [77].

Theorem 7.3.1. *Let* k *be a commutative field,* $m > 1$ *and* $c, d \geq 1$. *Then the following conditions are equivalent:*

(a) $E(k\langle X\rangle_c) \subseteq E(k\langle X\rangle_d)$, *i.e. every rational identity in* PI-*degree* d *is one for* PI-*degree* c,

(b) *there is an* X-*specialization* $k\langle X\rangle_d \longrightarrow k\langle X\rangle_c$,

(c) *there is a surjective local homomorphism* $D_d \longrightarrow D_c$, *where* D_i *is a division algebra over* k *of* PI-*degree* i,

(d) $c \mid d$.

Proof. (a) ⇒ (b) ⇒ (c) is clear. To prove (c) ⇒ (d), let $D'_d \subseteq D_d$ be a local ring with residue class field D_c, then $c = \text{PI-deg } D_c \mid \text{PI-deg } D'_d \mid \text{PI-deg } D_d$ by (i), (ii) of Th. 7A. (d) ⇒ (a): Let E be an infinite k-field. Since $c|d$, we can embed $\mathfrak{M}_c(E)$ in $\mathfrak{M}_d(E)$ by mapping α to $\text{diag}(\alpha, \alpha, \ldots, \alpha)$. Then every rational identity in $\mathfrak{M}_d(E)$ holds in $\mathfrak{M}_c(E)$. But these identities are just the rational relations in $k\langle X\rangle_d$, $k\langle X\rangle_c$, hence $E(k\langle X\rangle_c) \subseteq E(k\langle X\rangle_d)$, i.e. (a). ■

7.4 A special type of rational identity

As a consequence of Th.7.3.1 there are rational identities holding in PI-degree 3 but not in PI-degree 2. We shall now describe a particular example of such an identity which was found by Bergman [77]. From results in Bergman-Small [75] (cf.7.3) it follows that there is no (x,y)-

specialization

$$k\langle x,y\rangle_3 \longrightarrow k\langle x,y\rangle_2.$$

Thus there must be a relation holding in PI-degree 3 but not 2, and we are looking for an explicit such relation. We shall need some preparatory lemmas; we put $[X,Y] = XY-YX$.

Lemma 7.4.1. *Let* C *be a commutative ring and* $X,Y \in \mathfrak{M}_3(C)$, *then*

$$(1) \qquad [X,[X,Y]^2] = (\det[X,Y]).[X,[X,Y]^{-1}],$$

whenever $[X,Y]^{-1}$ *is defined. For* 2×2 *matrices the left-hand side of* (1) *is* 0.

Proof. Put $Z = [X,Y]$, then $\operatorname{tr} Z = 0$, hence Z has the characteristic equation $Z^3 + pZ - q = 0$, where $q = \det Z$. Now multiply by Z^{-1}: $Z^2 + p - qZ^{-1} = 0$; apply $[X,-]$: $[X,Z^2] - q[X,Z^{-1}] = 0$, i.e. (1).

For 2×2 matrices, $Z^2 - q = 0$, hence $[X,Z^2] = 0$. ■

If we write $Y' = [X,Y]$, the conclusion of the lemma can be expressed as

$$(2) \qquad \frac{((Y')^2)'}{((Y')^{-1})'} = \begin{cases} 0 & \text{for } 2 \times 2 \text{ matrices,} \\ \det Y' & \text{for } 3 \times 3 \text{ matrices.} \end{cases}$$

Here we have used the convention of writing $\frac{u}{v} = \alpha$ if $u = \alpha v$ for a scalar α.

Lemma 7.4.2. *Let* $X, Y \in \mathfrak{M}_3(C)$ *and write* Δ *for the discriminant of the characteristic polynomial of* X, *then*

$$(3) \qquad \det Y''' = \Delta \det Y'.$$

Proof. First let $X = \operatorname{diag}(\lambda_1,\lambda_2,\lambda_3)$ and $\delta = (\lambda_1 - \lambda_2)$

$(\lambda_2 - \lambda_3)(\lambda_3 - \lambda_1)$, then $\Delta = \delta^2$. Now an iterated commutator has the form

$$Y^{(n)} = \begin{pmatrix} 0 & (\lambda_1-\lambda_2)^n y_{12} & (\lambda_1-\lambda_3)^n y_{13} \\ (\lambda_2-\lambda_1)^n y_{21} & 0 & (\lambda_2-\lambda_3)^n y_{23} \\ (\lambda_3-\lambda_1)^n y_{31} & (\lambda_3-\lambda_2)^n y_{32} & 0 \end{pmatrix}$$

hence $\det Y^{(n)} = (\lambda_1-\lambda_2)^n(\lambda_2-\lambda_3)^n(\lambda_3-\lambda_1)^n y_{12}y_{23}y_{31} + (\lambda_1-\lambda_3)^n(\lambda_3-\lambda_2)^n(\lambda_2-\lambda_1)^n y_{13}y_{32}y_{21}$

$$= \delta^n(y_{12}y_{23}y_{31} + (-1)^n y_{13}y_{32}y_{21}).$$

Now $n = 1,3$ differ by a factor $\delta^2 = \Delta$, hence (3). This proves (3) for matrices over an algebraically closed field whenever $\Delta \neq 0$; hence it holds identically. ■

Using (2) and (3) we can write down rational identities for 3 x 3 matrices, but most of them will hold for 2 x 2 matrices too. What we need is a relation between determinants of commutators of 3 x 3 matrices which fails when these commutators are replaced by 0. For any X and Y let us again write $Y' = [X,Y]$ and consider

(4) $\det Y' \det Y''(\det(Y''^{-1})')(\det(Y'''^{-1})')$.

Since ' is a derivation, $\det(Y^{-1})' = \det - Y^{-1}Y'Y^{-1} = (\det Y)^{-2} \det - Y'$ and (4) becomes

$$(\det Y')(\det Y'')(\det Y'')^{-2}(\det Y''')(\det Y''')^{-2}(\det Y^{iv})$$

$$= (\det Y')(\det Y'')^{-1}(\det Y''')^{-1}(\det Y^{iv}).$$

Applying Lemma 7.4.2, we get

(5) $(\det Y')(\det Y'')^{-1}\Delta^{-1}(\det Y')^{-1}\Delta(\det Y'') = 1.$

Thus we obtain

Theorem 7.4.3. *(Bergman* [76]*). Let* k *be a commutative field and* n = 2 *or* 3; *for* $X, Y \in \mathfrak{M}_n(k)$ *write* $Y' = [X,Y]$, $\delta(Y) = (Y^2)'[(Y^{-1})']^{-1}$, *so that by* (2), $\delta(Y') = \det Y'$ *or* 0 *according as* n = 3 *or* 2, *then there are rational identities*

$$(6) \qquad \delta(Y')\delta(Y'')[(\delta(Y'')^{-1})'][(\delta(Y''')^{-1})'] = \begin{cases} 1 & \text{if } n = 3, \\ 0 & \text{if } n = 2. \end{cases}$$

Proof. By equating the left-hand side to 1 we get an identity in degree 3 but not in degree 2. We know that this holds if the left-hand side is defined, so we need only find X, Y for which the left-hand side is defined. Let K be any extension of k with more than two elements and write S for the set of matrices $\begin{pmatrix} 0 & a \\ b & 0 \end{pmatrix}$ $a, b \in K^*$ when n = 2, or

$$\begin{pmatrix} 0 & a & 0 \\ 0 & 0 & b \\ c & 0 & 0 \end{pmatrix} \qquad \begin{pmatrix} 0 & 0 & a \\ b & 0 & 0 \\ 0 & c & 0 \end{pmatrix} \qquad a, b, c \in K^*$$

when n = 3. Then S consists of invertible matrices and is closed under inversion and commutation by diagonal matrices with distinct elements. If we choose Y in S and X diagonal with distinct entries, all terms lie in S and so (6) is defined. ∎

7.5 The rational meet of a family of X-rings

We shall now make a closer study of specializations, following Bergman [76]. We shall find that for skew fields they cannot be reduced to the situation involving only two fields, as in the commutative case. We shall be concerned with two basic notions: an *essential term* in a family of X-fields and the *support relation*.

Given rings $R_1 \subseteq R_2$ we say that R_1 is *rationally closed* in R_2 if the inclusion is a local homomorphism. The intersection of a rationally closed family is again rationally closed, so we can speak of the *rational closure* of X in R, which is the least rationally closed subring of R containing X. If it is R, we call R a *strict X-ring*; e.g. $\mathbf{Q}(x,y)$ is a strict (x,y)-ring, so is $\mathbf{Z}[x,y,y^{-1}]$, but not $\mathbf{Z}[x,y,xy^{-1}]$. Generally, if Σ is the set of all matrices over $\mathbf{Z}\langle X\rangle$ which are mapped to invertible matrices over R, then the rational closure of X in R is contained in the Σ-rational closure of $\mathbf{Z}\langle X\rangle$, in the sense of Ch.4, but the two may be distinct (if $x,y,u,v \in X$, the entries of $\begin{pmatrix} x & y \\ u & v \end{pmatrix}^{-1}$ lie in the latter but not the former). We note that an epic $\mathbf{Z}\langle X\rangle$-field, briefly an epic X-field, is just a strict X-field.

A local homomorphism between X-fields $\phi: D \longrightarrow D'$ may be described as a partial homomorphism from D to D' whose graph is rationally closed in $D \times D'$; hence if there is any X-specialization at all, the rational closure of X in $D \times D'$ is the unique least X-specialization. So there is at most one minimal X-specialization between two X-fields. Our aim is to study the rational closure of X in finite direct products; to do so we need to introduce the following basic concepts.

Definition. Let $\{R_s\}_S$ be a family of strict X-rings, then their *rational meet* $\bigwedge_S R_s$ is the rational closure of X in $\prod_S R_s$.

The rational meet can also be viewed as the product in the category of strict X-rings. We note: the bigger S is, the smaller is $\bigwedge_S R_s$, in the sense that for $T \subseteq S$ we have a *projection* $p_{ST}: \bigwedge_S R_s \longrightarrow \bigwedge_T R_s$. E.g. whether $D_1 \wedge D_2$ is the graph of a specialization in one direction or the other depends on which projection maps are injective.

Lemma 7.5.1. *Let* $\{R_s\}_S$ *be any family of strict X-rings, then*

$$E(\bigwedge_S R_s) = \bigcap_S E(R_s), \qquad Z(\bigwedge_S R_s) = \bigcap_S Z(R_s).$$

For $\bigwedge_S R_s$ is the set of all rational expressions evaluable in each R_s modulo the relation of having equal values in each R_s: $f \sim g \iff f^{R_s} = g^{R_s}$ for all $s \in S$.■

Let $\{D_s\}_S$ be a finite family of epic X-fields and let $t \in S$. We shall call the index t (also the field D_t) *essential* in S if there exists $f \in \bigcap_S E(D_s)$ such that $f \in Z(D_t)$, $f \notin Z(D_s)$ for all $s \neq t$. Equivalent formulations are

(1) $E(D_t) \not\supseteq \bigcap_{s \neq t} E(D_s)$,

or writing Ker D for the singular kernel of ϕ: $\mathbf{Z}\langle X\rangle \longrightarrow D$,

(2) $\mathrm{Ker}(D_t) \not\subseteq \bigcup_{s \neq t} \mathrm{Ker}(D_s)$.

For when (1) holds, take $f \in \bigcap E(D_s)$, $f \notin E(D_t)$, then f contains a subexpression g^{-1} such that $g^{D_t} = 0$ but $g^{D_s} \neq 0$ for $s \neq t$. Conversely, given such g, we find that g^{-1} belongs to the right but not the left-hand side of (1); the equivalence of (1) and (2) is clear. Using this notion, we can say when the rational meet reduces to a direct product:

Proposition 7.5.2. *Let* X *be a set and* $\{D_s\}_S$ *a finite family of epic* X*-fields, then the following are equivalent:*

(a) *Each* s *is essential in* S*,*

(b) *for each* $s \in S$*, there exists* $e_s \in \bigcap_S E(D_u)$ *such that* $e_s^{D_t} = \delta_{st}$*,*

(c) $\bigwedge_S D_s = \prod_S D_s$.

Proof. (a) $\Rightarrow$ (b). Choose f_s defined in all D_t and vanishing in D_s but not in D_t for $t \neq s$. Then $g_t = \prod_{s \neq t} f_s$ (in any order) vanishes on all D's except D_t. Now $e_s = g_s(\Sigma_t g_t)^{-1}$ satisfies the required condition.

(b) $\Rightarrow$ (c). By (b), $\bigwedge_S D_s$ contains a set of central idempotents e_s, which shows that $\bigwedge_S D_s = \prod_S R_s$ for some $R_s \subseteq D_s$. Now $\wedge D_s$ is rationally closed in $\prod D_s$, hence R_s is rationally closed in D_s and it contains X, so $R_s = D_s$.

(c) $\Rightarrow$ (a). Given $s \in S$, choose $g \in \cap E(D_t)$ such that $g^{D_s} = 0$ but $g^{D_t} \neq 0$ for $t \neq s$. ■

Illustration. Consider $D_1 \wedge D_2$; if $E(D_1) \supseteq E(D_2)$, $D_1 \wedge D_2$ is a local ring, the graph of a specialization $D_1 \longrightarrow D_2$. Similarly if $E(D_1) \subseteq E(D_2)$, while if neither inclusion holds, then $D_1 \wedge D_2 = D_1 \times D_2$. For more than two factors we shall see that $\wedge D_s$ is a *semilocal* ring, i.e. a ring R such that $R/J(R)$ is semisimple Artinian.

Lemma 7.5.3. *Let* $f:R \longrightarrow R'$ *be a homomorphism such that* Rf *rationally generates* R', *then* f *is local if and only if* f *is surjective and* $\ker f \subseteq J(R)$.

Proof. $\Rightarrow$. Rf is rationally closed because f is local and it rationally generates R', hence Rf = R'. If af = 0, then 1 + ax maps to 1, a unit, hence 1 + ax is a unit, for any $x \in R$, but that means that $a \in J(R)$.

$\Leftarrow$. If the conditions hold, take $a \in R$ such that af is a unit, say af.bf = 1 for some $b \in R$. Then ab = 1 + n, $n \in J(R)$, hence $ab(1 + n)^{-1} = 1$ and similarly $(1 + m)^{-1}ba = 1$ for some $m \in J(R)$, so a is a unit. ■

Note that the extra hypothesis (Rf rationally generates R') is needed only for $\Rightarrow$, not for $\Leftarrow$.

We can now prove a result which describes the structure of rational meets:

Proposition 7.5.4. *Let* X *be a set,* $\{D_s\}_S$ *a finite family of epic* X*-fields, pairwise non-isomorphic as* X*-fields, and* $S_o \subseteq S$, *then the following conditions are equivalent:*

(a) *The projection* $p: \bigwedge_S D_s \longrightarrow \bigwedge_{S_o} D_s$ *is a local homomorphism,*

(b) $p\colon \bigwedge_S D_s \longrightarrow \bigwedge_{S_o} D_s$ *is surjective, with kernel* $\subseteq J(\bigwedge_S D_s)$,

(c) $\bigcap_S E(D_s) = \bigcap_{S_o} E(D_s)$,

(d) S_o *includes all essential indices in* S.

Further, let $U \subseteq S$ *be the subset of all essential indices, then* $\bigwedge_U D_s = \prod_U D_s$ *and* $\bigwedge_S D_s$ *is a semilocal ring with residue class ring* $\prod_U D_s$:

$$(3) \qquad \bigwedge_S D_s / J(\bigwedge_S D_s) \cong \prod_U D_s.$$

Proof. (a) $\iff$ (b) by the lemma and (a) $\iff E(\bigwedge_S D_s) = E(\bigwedge_{S_o} D_s)$, which is equivalent to (c). Now let $V \subseteq S$ be a subset which is minimal subject to (a)-(c). By (c), V contains no inessential suffix, i.e. $V \subseteq U$, hence $\bigwedge_V D_s = \prod_V D_s$ by Prop. 7.5.2. This ring is semisimple and V satisfies (b), so $\bigwedge_S D_s / J(\bigwedge_S D_s) \cong \bigwedge_V D_s = \prod_V D_s$, hence $\bigwedge_S D_s$ is semilocal, with residue-class fields isomorphic to the D_s $(s \in V)$. But distinct D's are non-isomorphic as X-fields, so V is the unique minimal subset of S satisfying (a)-(c). Therefore (a)-(c) are equivalent to

(d') $S_o \supseteq V$.

Now for any $t \in S$, $t \notin U \iff S\setminus\{t\}$ satisfies (c) $\iff S\setminus\{t\} \supseteq V$ because (c)$\iff$(d'))$\iff t \in V$. Hence $V = U$ as claimed

By (3) we have

Corollary 1. *The set* U *of essential suffixes in* S *can also be characterized as the set of those* $t \in S$ *for which* $p_{S\,t}\colon \bigwedge_S D_s \longrightarrow D_t$ *is surjective.* ■

The complement of U, the set of *inessential* suffixes is the set of $t \in S$ such that (a) $p_{S\;S\setminus\{t\}}$ is a local homomorphism, (b) $p_{S\;S\setminus\{t\}}$ is surjective with kernel in J, or equivalently (c) $\bigcap_S E(D_s) = \bigcap_{S\setminus\{t\}} E(D_s)$. This follows from Prop.7.5.4. Here (c) states essentially that $\mathrm{Ker}(D_t) \subseteq \cup_S \mathrm{Ker}(D_s)$. In the commutative case this can happen only when $\ker(D_t) \subseteq \ker(D_s)$ for some s, i.e.

Corollary 2. *In the case of a family* $\{D_s\}_S$ *of commutative fields a suffix* t *is inessential in* S *if and only if* D_t *has some* D_s $(s \neq t)$ *as specialization.* ■

To give an example, let $S = \{0,1,2\}$ and suppose that 0 is inessential in S, then every relation holding in D_o holds in D_1 or D_2. In general Cor.2 need not hold, i.e. there may be f_1, f_2 such that $f_1 = 0$ in D_o but not in D_1 and $f_2 = 0$ in D_o but not in D_2. Now $f_1 + f_2$ would seem to be 0 in D_o but not in D_1 or D_2, and it need not be defined ($f_1 \neq 0$ may hold 'degenerately' in D_1, i.e. f_1 is not defined in D_1). This will become clear later.

We now come to the second basic notion, the support relation. We have seen that $p_{S\,t}: \bigwedge_S D_s \longrightarrow D_t$ is surjective if and only if t is essential in S. Our next question is: When is $p_{S\,t}$ injective? This is answered by

Proposition 7.5.5. *Let* X *be a set,* $\{D_s\}_S$ *a family of epic* X-*fields and* $t \in S$, *then the following are equivalent*:

(a) $Z(D_t) \cap \bigcap_S E(D_s) \subseteq \bigcap_S Z(D_s)$,

(b) *any relation defined in each* D_s *and holding in* D_t *holds in all* D_s,

(c) $p_{S\,t}: \bigwedge_S D_s \longrightarrow D_t$ *is injective.*

Note that by (c) there is a local homomorphism $D_t \longrightarrow \prod_S D_s$.

Proof. $e \in R(X)$ represents an element in $\ker p_{S\,t}$ if and only if it is in the left-hand side of (a) and it represents 0 if and only if it is in the right-hand side of (a); this just expresses (b). ■

When these conditions hold we say that t *supports* S (or also: D_t supports $\{D_s\}_S$). More generally, if $t \notin S$, we say that t supports S if it supports $S\cup\{t\}$ in the above sense.

To gain an understanding of the support relation we begin by proving some trivial facts.

Proposition 7.5.6. *Let* X *be a set and* $\{D_s\}_S$ *a family of epi*[illegible] X*-fields. Then*

(i) *Let* $t \in S$, $U \subseteq S$. *If* t *supports* U *and* U *contains an element distinct from* t, *then* t *is essential for* $U\cup\{t\}$.

(ii) *If* t *supports* U, *then it supports* $U\cup\{t\}$.

(iii) *If* t *supports* S_i $(i \in I)$, *then it supports* $\bigcup_I S_i$.

(iv) *If* t *supports* U *and for each* $u \in U$, $u \in S_u$ *and* u *supports* S_u, *then* t *supports* $\bigcup_U S_u$.

Proof. 't is inessential for S' means: any relation defined in all D_s and holding in D_t also holds in *some* D_s, $s \neq t$. 't supports S' means: any relation defined in all D_s and D_t and holding in D_t holds in *all* D_s. Now (i) is clear and (ii) also follows. To prove (iii), let f be defined in D_t and D_s $(s \in S_i)$ and $f = 0$ in D_t, then $f = 0$ in all D_s, $s \in S_i$ so t supports $\cup S_i$. (iv) Let f be defined in D_t and D_v where $v \in S_u$, for all $u \in U$. If $f = 0$ in D_t then $f = 0$ in D_u $(u \in$ hence $f = 0$ in D_v $(v \in S_u)$, so t supports $\bigcup_U S_u$. ■

Corollary. If the D_s *are commutative and* t *supports* S, *then either* $S = \emptyset$ *or* D_t *specializes to some* D_s $(s \in S)$. *More precisely:* t *supports* {s} *if and only if there is a specialization* $D_t \longrightarrow D_s$.

For if t supports S and $S \neq \emptyset$, then either $t \in S$ or t is inessential for $S\cup\{t\}$; in the latter case there exists $s \neq t$ in S such that D_s is a specialization of D_t. ■

To clarify the relation between support and essential set we have the following lemma. Note that by (i) above, a supporting index is a special kind of inessential index.

Lemma 7.5.7. *Let* X *be a set,* $\{D_s\}_S$ *a finite family of pairwise non-isomorphic epic* X*-fields and* D_t *an epic* X*-field. Then the following are equivalent:*

(a) $S_\cup\{t\}$ *is a minimal set in which* t *is inessential,*

(b) S *is a minimal non-empty set supported by* t.

Proof. Let us write (a_o),(b_o) for (a),(b) without the minimality clause. Then $(b_o) \Rightarrow (a_o)$ by (i) of Prop.7.5.6. To prove that (a) $\Rightarrow$ (b_o) we know by hypothesis that S is minimal subject to $\bigcap_S E(D_s) = \bigcap_{S\cup\{t\}} E(D_s)$. By Prop. 7.5.4, S is the set of essential indices of $S_\cup\{t\}$, hence the projection $p_{S_\cup\{t\}S}$ is surjective. If t does *not* support S, there exists $a \in \bigwedge_{S\cup\{t\}} D_s$ and $u \in S$ such that $a^{D_t} = 0$ but $a^{D_u} \neq 0$. Let us write a_u for a^{D_u} etc. Since the map $\bigwedge_{S\cup\{t\}} D_s \longrightarrow \prod_S D_s$ is surjective, there exists $b \in \bigwedge_{S\cup\{t\}} D_s$ such that $b_u = a_u^{-1}$, $b_s = 0$ for all $s \neq u$, t, where $s \in S$. Then $e = ab$ is in $\bigwedge_{S\cup\{t\}} D_s$ and has value 1 in D_u and 0 everywhere else, for $b_s = 0$ for $s \neq t$ and $a_t = 0$. Thus e is a central idempotent and so $\bigwedge_{S\cup\{t\}} D_s = R \times D_u$. Now write $S^* = (S\setminus\{u\})\cup\{t\}$, then $R \subseteq \prod_{S^*} D_s$ and R is rationally generated by X and rationally closed, hence $R = \bigwedge_{S^*} D_s$. Further, $p_{S\cup\{t\}S}$ is a local homomorphism, so $p_{S^*\ S\setminus\{u\}}$ is too (we have to factor by D_u), therefore by Prop.7.5.4, $S\setminus\{u\}$ includes all essential indices in S, which contradicts the minimality of S. So D_t supports $\{D_s\}_S$ and (b_o) follows.

Thus we have (a) $\Rightarrow$ (b_o) and (b) $\Rightarrow$ (a_o); in an obvious terminology, if S is a minimal a-set, it is a b-set. Now take a minimal b-subset S' of S; this is also an a-set contained in S, hence S' = S, i.e. S was a minimal b-set. Thus (a) $\Rightarrow$ (b) and similarly (b) $\Rightarrow$ (a). ■

Corollary 1. $E(D_t) \supseteq \bigcap_S E(D_s)$ *if and only if* D_t *supports some non-empty subfamily of* $\{D_s\}_S$.

For the left-hand side expresses the fact that t is inessential in $S\cup\{t\}$. Now pick $S_o \subseteq S$ minimal with this property and apply the lemma to obtain the desired conclusion.■

A relation 't supports S' will be called *non-trivial* if $S \neq \emptyset, \{t\}$.

Corollary 2. Each $s \in S$ *is essential if and only if there are no non-trivial support relations in* S.■

The essential relations are determined by the minimal essential relations, but there is no corresponding statement for support relations. However, Cor.2 shows that essential relations are determined by the support relations.

Let us call a set S *essential* if each member is essential in it.

Proposition 7.5.8. Let $\{D_s\}_{S\cup\{t\}}$ *be a finite family of pairwise non-isomorphic epic* X*-fields, then the following conditions are equivalent:*

(a) t *supports* S *and* S *is essential,*

(b) $\bigwedge_{S\cup\{t\}} D_s$ *is a semilocal ring contained in* D_t *(via the projection map), with residue class fields* D_s*,*

(c) *there exists a semilocal* X*-ring* $R \subseteq D_t$ *with residue class fields* D_s $(s \in S)$*,*

(d) $Z(D_t) \cap \bigcap_S E(D_s) \subseteq \bigcap_S Z(D_s)$ *and no* $E(D_s)$ *contains the intersection of all the others.*

Proof. (a) $\Rightarrow$ (b) by Prop.7.5.4,5, (b) $\Rightarrow$ (c) $\Rightarrow$ (d) is trivial and (d) $\Rightarrow$ (a) is also clear.■

Corollary. Let $\{D_s\}_S$ *be a finite family of pairwise non-isomorphic epic* X*-fields,* $t \in S$ *and suppose that* t *supports* S *and* U *is the subset of essential indices, then the map* $\bigwedge_S D_s \longrightarrow \bigwedge_{U\cup\{t\}} D_s$ *is an isomorphism and* t *supports* U.■

7.6 The support relation

We shall now give a complete description of all possible support relations, using the work of Bergman-Small [75] (and still following Bergman [76]). We shall need Th.6.8 of that paper, which for our purpose may be stated as follows.

Theorem 7.B *Let* R *be a prime* PI-*ring and* $\mathfrak{p}_1$ *a prime ideal of* R, *then* PI-deg R - PI-deg R/$\mathfrak{p}_1$ *can be written as a sum of integers* PI-deg R/$\mathfrak{p}$ *(allowing repetitions), where* $\mathfrak{p}$ *ranges over the maximal ideals of* R.

Let us say that an integer n *supports* a set M of positive integers if for each $m \in M$, n-m lies in the additive monoid generated by the elements of M. Clearly M must then be a subset of $\{1,2,\ldots,n\}$. The Bergman-Small theorem shows the truth of the following:

If R *is any prime* PI-*ring, then* PI-deg R *supports the set* $\{\text{PI-deg } R/\mathfrak{p} \mid \mathfrak{p} \text{ prime in } R\}$.

In what follows, X will be fixed, with more than one element, so that $k\langle X\rangle_n$ has PI-degree n. We shall write E(n) = $E(k\langle X\rangle_n)$, Z(n) = $Z(k\langle X\rangle_n)$ for brevity.

Theorem 7.6.1. *(Bergman* [76]*). Let* n *be a positive integer and* M *a finite (non-empty) set of positive integers. Then the following conditions are equivalent:*

(a) $k\langle X\rangle_n$ *supports* $\{k\langle X\rangle_m \mid m \in M\}$,

(b) $Z(n) \cap \bigcap_M E(m) \subseteq \bigcap_M Z(m)$,

(c) $p_{M\cup\{n\}\ \{n\}}: \widehat{\prod_{M\cup\{n\}}} k\langle X\rangle_i \longrightarrow k\langle X\rangle_n$ *is injective,*

(d) *there exists a prime* PI-*ring* R *of* PI-degree n *such that the set of* PI-degrees *of the residue class rings of* R *at maximal ideals is precisely* M, *briefly:*
PI-deg(R/max) = M.

(e) n *supports* M.

Proof. (a)-(c) are equivalent by Prop.7.5.5. Now we have

two variants of (d), one weaker and one stronger:

(d−) There is a prime ring R of PI-degree n such that

$$\text{PI-deg}\{R/\text{prime}\} \supseteq M \supseteq \text{PI-deg}\{R/\max\},$$

(d+) there is a semilocal prime ring R of PI-degree n such that every non-zero prime ideal is maximal and PI-deg{R/max} = M, and every residue class field is infinite.

We complete the proof by showing (c) $\Rightarrow$ (d−) $\Rightarrow$ (e) $\Rightarrow$ (d+) $\Rightarrow$ (b). Note that each condition implies that $M \subseteq \{1,2,\ldots,n\}$. (c) $\Rightarrow$ (d−). Put $R = \bigwedge_{M \cup \{n\}} k \langle X \rangle_i$; by Prop. 7.5.4, the residue class rings at the maximal ideals are among the $k \langle X \rangle_m$ ($m \in M$), for n itself cannot occur, by (c) and Prop. 7.5.6. Now for each $m \in M$, $\ker(R \longrightarrow k\langle X \rangle_m) = \mathfrak{p}$ is prime and $\text{PI-deg}(R/\mathfrak{p}) = m$, hence (d−).

(d−) $\Rightarrow$ (e) is just the Bergman-Small theorem quoted earlier.

(e) $\Rightarrow$ (d+). By (e) we can write $n = m(1,1)+\ldots+m(1,r_1) = \ldots = m(s,1)+\ldots+m(s,r_s)$, $m(i,j) \in M$, where $s \geq 1$, each $r_i \geq 1$ and each $m \in M$ occurs as some $m(i,j)$. Let A be a commutative k-algebra which is a semilocal principal ideal domain with just s non-zero prime ideals $\mathfrak{I}_1,\ldots,\mathfrak{I}_s$ each with infinite residue class field $K_i = A/\mathfrak{I}_i$ (e.g. let $K \supseteq k$ be an infinite field extension and take a suitable localization of $K[t]$). Then $A/J(A) = \prod_i K_i$, hence $\mathfrak{M}_n(A)/J(\mathfrak{M}_n(A)) \cong \prod_i \mathfrak{M}_n(K_i)$. Now for each i, $\mathfrak{M}_n(K_i)$ has a block diagonal subring isomorphic to

$$\mathfrak{M}_{m(i,1)}(K_i) \times \ldots \times \mathfrak{M}_{m(i,r_i)}(K_i) = L_i \qquad \text{say.}$$

Hence

$$(1) \quad Q = \prod_i L_i \subseteq \prod_i \mathfrak{M}_n(K_i) \cong \mathfrak{M}_n(A)/J(\mathfrak{M}_n(A)),$$

where Q as a direct product of simple Artinian rings is semisimple. Let R be the inverse image of Q in $\mathfrak{M}_n(A)$, by the isomorphism (1), then $J(R) = J(\mathfrak{M}_n(A))$, hence $R/J(R) \cong Q$. Since $R/J(R)$ is semisimple (Artinian), it follows that R is semilocal and PI-deg{R/max} = M. Let $\mathfrak{P}$ be a prime ideal in R and $\mathfrak{p} = \mathfrak{P} \cap A$, then $\mathfrak{p}$ is prime in A, so $\mathfrak{p}$ is 0 or some $\mathfrak{I}_i$. Suppose that $\mathfrak{p} = 0$, then $A \subseteq R/\mathfrak{P}$; write F for the field of fractions of A, then since $A + \mathfrak{M}_n(J(A)) \subseteq R \subseteq \mathfrak{M}_n(A)$, we have $R_{A*} = \mathfrak{M}_n(A)_{A*} = \mathfrak{M}_n(F)$ because A is a domain and $J(A) \neq 0$. Hence R_{A*} is simple with 0 as the only prime, so $\mathfrak{P}$ must be 0. If $\mathfrak{p} = \mathfrak{I}_i$, then $K_i = A/\mathfrak{I}_i \subseteq R/\mathfrak{P}$ and since R is a finitely generated A-module, $R/\mathfrak{P}$ is a finitely generated K_i-module, hence Artinian. It is also prime, hence simple, and so $\mathfrak{P}$ was maximal. Thus R satisfies (d+).

(d+) $\Rightarrow$ (b). Assume that e lies in the left-hand side of (b), i.e. e = 0 is a rational identity holding in PI-degree n and not degenerate in PI-degree m for any $m \in M$. We have to show that e = 0 holds in each PI-degree $m \in M$. Let R be as in (d+); this means that for each prime $\mathfrak{P} \neq 0$ of R there is given a map $\alpha_{\mathfrak{P}}: X \longrightarrow R/\mathfrak{P}$ such that $e^{\alpha\mathfrak{P}}$ is defined in $R/\mathfrak{P}$; to show that all the $e^{\alpha\mathfrak{P}}$ are 0. Since R is semilocal, by the Chinese remainder theorem there exists $\alpha: X \longrightarrow R$ inducing all the $\alpha_{\mathfrak{P}}$. Now e^{α} can be evaluated (mod $\mathfrak{P}$) for all maximal $\mathfrak{P}$, hence it can be evaluated in R. Since $e \in Z(n)$, $e^{\alpha} = 0$ and so $e^{\alpha\mathfrak{P}} = 0$ as claimed. ■

To give an illustration, we have 5 = 2 + 3. Let A be a local principal ideal domain with maximal ideal $\mathfrak{I}$, then $\mathfrak{M}_5(A)$ contains the subring

$$R = \begin{pmatrix} A_2 & {}^2\mathfrak{I}^3 \\ {}^3\mathfrak{I}^2 & A_3 \end{pmatrix}$$

and we have a local homomorphism $\mathfrak{M}_5(A) \longrightarrow \mathfrak{M}_2(K) \times \mathfrak{M}_3(K)$ ($K = A/\mathfrak{I}$). This gives rise to a specialization of fields, because when we replace $\mathfrak{M}_n(K)$ by the generic matrix ring, we get a field with the same identities as $\mathfrak{M}_n(K)$.

If we combine Th.7.6.1 with Prop. 7.5.8, we get

Corollary 1. *Let* n *be an integer and* M *a set of integers, then the following conditions are equivalent:*

(a) $p: \widehat{{}_{M\cup\{n\}}} k \langle X \rangle_i \longrightarrow k \langle X \rangle_n$ *is injective, with residue class fields* $k \langle X \rangle_m$ $(m \in M)$,

(b) $k \langle X \rangle_n$ *has a semilocal subring with residue class fields* $k \langle X \rangle_m$ $(m \in M)$,

(c) M *is a minimal set supported by* n. ■

Corollary 2. *Let* n, M *be as before, then the following are equivalent:*

(a) $k \langle X \rangle_n$ *supports a non-empty subfamily of* $\{k \langle X \rangle_m \mid m \in M\}$,

(b) *every rational identity holding in* PI-degree n *holds in some* PI-degree m $(m \in M)$: $Z(n) \cap \bigcap_M E(m) \subseteq \bigcup_M Z(m)$.

(c) *there exists a prime ring of* PI-degree n *with* {PI-deg R/max} $\subseteq$ M,

(d) n *supports a subset of* M. ■

To describe the connexion between prime ideals and the support relation we shall need a couple of auxiliary lemmas.

Lemma 7.6.2. *Let* R *be a ring,* $\mathfrak{I}_1,\ldots,\mathfrak{I}_m$ *any ideals in* R *and* $\mathfrak{P}_1,\ldots,\mathfrak{P}_n$ *any prime ideals such that* $\mathfrak{I}_i \not\subseteq \mathfrak{P}_j$ *for all* i,j, *then* $\cap \mathfrak{I}_i \not\subseteq \cup \mathfrak{P}_j$.

Proof. If $\mathfrak{P}_{j'} \subseteq \mathfrak{P}_j$, we can omit $\mathfrak{P}_{j'}$. Since $\mathfrak{P}_j$ is prime, we then have $\mathfrak{I}_1\ldots\mathfrak{I}_m\mathfrak{P}_1\ldots\mathfrak{P}_{j-1}\mathfrak{P}_{j+1}\ldots\mathfrak{P}_n \not\subseteq \mathfrak{P}_j$. Choose a_j in the left but not the right-hand side, then $a = a_1 + \ldots + a_n$

$\in \mathfrak{I}_i$ for all i but $a \notin \mathfrak{P}_j$. ■

Throughout, $\{D_s\}_S$ is a finite family of epic X-fields, $R = \bigwedge_S D_s$ and $\mathfrak{P}_s = \ker(R \longrightarrow D_s)$. Thus $\mathfrak{P}_s = 0$ if and only if s supports S and $\mathfrak{P}_s$ is maximal if and only if s is essential. We shall write Ess(S) for the set of essential suffixes in S.

Lemma 7.6.3. *Assume that the* D_s *are pairwise non-isomorphic as* X*-fields, and that* $T \subseteq S$, *then the following conditions are equivalent:*

(a) $\bigwedge_S D_s \longrightarrow \bigwedge_T D_s$ *is surjective,*

(b) *for each* $t \in T$ *and* $s \in \mathrm{Ess}(S)$, $\mathfrak{P}_t \subseteq \mathfrak{P}_s \Rightarrow s \in T$.

When these conditions hold, then $\mathrm{Ess}(T) = \mathrm{Ess}(S) \cap T$.

Proof. (a) ⇒ (b). Put $R' = \bigwedge_T D_s$, then $p:R \longrightarrow R'$ is surjective by hypothesis. If t,t' are as in (b), then $\mathfrak{P}_{t'}$ is a maximal ideal of R containing $\mathfrak{P}_t$ and hence ker p, so its image under p is a maximal ideal of R' with the same residue class field. But if R' has a residue class field isomorphic to $D_{t'}$, then $t' \in \mathrm{Ess}(T) \subseteq T$, so (b) holds.

(b) ⇒ (a). Since im p rationally generates R', it is enough to show that im p is rationally closed in R', i.e. the inclusion im $p \subseteq R'$ is a local homomorphism. Let $a \in R$ be such that ap is invertible in R', then $a \notin \mathfrak{P}_t$ for all $t \in T$. Now consider those $s \in \mathrm{Ess}(S)$ for which $a \in \mathfrak{P}_s$; by (b), since $s \notin T$, $\mathfrak{P}_s \not\supseteq \mathfrak{P}_t$ for all $t \in T$. Hence $\mathfrak{P}_s \not\supseteq \bigcap_T \mathfrak{P}_t = \ker p$. By Lemma 7.6.2 there exists $b \in R$ such that $b \in \ker p$ and for any $s \in \mathrm{Ess}(S)$, $b \notin \mathfrak{P}_s$ if and only if $a \in \mathfrak{P}_s$. But then a + b lies in no maximal ideal of R and so is a unit, and (a + b)p = ap is likewise a unit in im p. ■

Note that (a) shows that the residue class rings of R' at maximal ideals are just the residue class rings of R at the maximal ideals containing ker p. We can now express the inclusion of prime ideals in terms of the support relation. We shall write $\mathrm{Supp}_S(t)$ for the maximal subset of S supported by t, i.e. the union of all subsets supported by t.

Theorem 7.6.4. *Let* $\{D_s\}_S$ *be a finite family of epic* X-*fields,* $R = \bigwedge_S D_s$ *and* $\mathfrak{P}_s = \ker(R \longrightarrow D_s)$, *then*

$$\text{(2)} \qquad \mathrm{Supp}_S(u) = \{ v \in S \mid \mathfrak{P}_u \subseteq \mathfrak{P}_v \}.$$

Proof. Isomorphic X-fields determine the same kernel in R, so we may without loss of generality take the D_s to be pairwise non-isomorphic. Fix u and let T be the right-hand side of (2), i.e. the set of all $s \in S$ for which $\mathfrak{P}_s \supseteq \mathfrak{P}_u$, then T satisfies (b) of Lemma 7.6.3, so $R \longrightarrow \bigwedge_T D_s$ is surjective. The kernel is $\bigcap_T \mathfrak{P}_t$ which contains $\mathfrak{P}_u$ by definition of T, in fact since $u \in T$, we have $\bigcap_T \mathfrak{P}_s = \mathfrak{P}_u$. Hence the map $\bigwedge_T D_s \longrightarrow D_u$ is injective, i.e. u supports T. It follows that $T \subseteq \mathrm{Supp}_S(u)$, but clearly also $\mathrm{Supp}_S(u) \subseteq T$, so $\mathrm{Supp}_S(u) = T$. ■

Corollary 1. *In Lemma* 7.6.3, (b) *just states that* $T \supseteq \mathrm{Ess}\,(S) \cap \mathrm{eg} \bigcup_T \mathrm{Supp}_S(t)$. ■

Corollary 2. *If* $T, T' \subseteq S$, *then* $\ker p_{ST} \subseteq \ker p_{ST'}$ *if and only if* $\bigcup_T \mathrm{Supp}_S(t) \supseteq \bigcup_{T'} \mathrm{Supp}_S(t)$. *In particular,*

(i) p_{ST} *is injective* $\Longleftrightarrow \bigcup_T \mathrm{Supp}_S(t) = S$,

(ii) $\ker p_{ST} \subseteq J(R) \Longleftrightarrow \bigcup_T \mathrm{Supp}_S(t) \supseteq \mathrm{Ess}(S)$. ■

In general $R = \bigwedge_S D_s$ will have prime ideals not of the form $\mathfrak{P}_s$ e.g. if C is a commutative local domain, X is a rational generating set and D_o, D_1 are the field of fractions and the residue class field respectively, then $D_o \wedge D_1 = C$, but C may have other primes (if C is not discrete).

We recall that Prop. 7.5.2 asserted that $\bigwedge_S D_s = \prod_S D_s$ if and only if S is essential. More generally we can now say

$$\bigwedge_S D_s = \prod_i (\bigwedge_{S_i} D_s), \quad S = S_1 \cup \ldots \cup S_r$$

if the S_i are disjoint support sets in S (i.e. for any $t \in S_i$, $\mathrm{Supp}_S(t) \subseteq S_i$).

We know that non-isomorphic X-fields may have the same kernels, e.g. $k\langle x,y\rangle \subseteq k[t][x;\alpha_i]$, where $\alpha_i : f(t) \longmapsto f(t^i)$. Here $y = xt$ and $tx = xt^i$ (cf. page 15f.). The resulting embeddings $k\langle x,y\rangle \longrightarrow D_i$ are distinct for $i = 2,3,\ldots$ and none is a specialization of the others.

By contrast, if R is a right Ore X-ring, the R-fields may be determined by their kernels, e.g. $R = k\langle X\rangle_n$. If moreover, D_t is commutative, then $\mathrm{Supp}_S(t) = \{u \in S \mid t \text{ supports } \{u\}\}$, i.e. the set of u such that $D_t \longrightarrow D_u$ is a specialization. For let $C = \widehat{\mathrm{Supp}_S(t)}\, D_s$; we have an injection $C \longrightarrow D_t$, so C is a commutative integral domain with D_t as field of fractions. Let $u \in \mathrm{Supp}_S(t)$, then $C/\mathfrak{P}_u$ is an integral domain with fields of fractions D_u. Hence the localization at $\mathfrak{P}_u$ is a local ring $L_u \subseteq D_t$ with residue class ring D_u, i.e. we have a specialization $D_t \longrightarrow D_u$.

7.7 Examples

Before constructing examples let us summarize the properties of supports. This is most easily done by introducing the notion of an *abstract support system*. By this we understand a set S with a relation on S x P(S) written $t \propto U$ and called the *support relation*, with the following properties:

S.1 *If* $t \in S$, $U \subseteq S$, *then* $t \propto U \Longleftrightarrow t \propto U \cup \{t\}$,

S.2 *if* $t \propto S_i$ $(i \in I)$, *then* $t \propto \bigcup_I S_i$,

S.3 *if* $t \propto U$ *and for each* $u \in U$, $u \propto S_u \neq \emptyset$, *then* $t \propto \bigcup S_u \cup U$.

If in S.2 we take the index set I to be empty, the hypothesis is vacuous, hence $t \propto \emptyset$ and by S.1, $t \propto \{t\}$ always.

A special case of the support relation is that where

$$t \propto U \iff t \propto \{u\} \text{ for all } u \in U.$$

This is completely determined by all pairs t,u with $t \propto \{u\}$ and if we write $t \leq u$ instead of $t \propto \{u\}$ we obtain a preordering of S. Conversely, every preorder on S leads to a support relation in this way. Thus preorders may be regarded as a special case of support relations.

A support relation on S induces a support relation on any subset of S. If a support relation is such that

$$s \propto \{t\} \text{ and } t \propto \{s\} \text{ imply } s = t,$$

the relation is said to be *separated*. E.g. the separated preorders are just the partial orders. Note that this is *not* the same as $s \in \mathrm{Supp}_S(t)$, $t \in \mathrm{Supp}_S(s)$, which may well hold for distinct s,t in a separated support relation.

We now construct all possible separated support relations on a 3-element set. There are 10 in all, 5 of them orders (if we allow non-separated ones and do not identify isomorphic ones, we get 53 support systems, 29 of them preorders).

We list the 10 below, the orders first, with rising arrows to indicate specializations.

1. $\circ D_1 \quad \circ D_2 \quad \circ D_3$ — Examples. (a) $X=\emptyset$, $D_i = \mathbf{Z}/p_i$ ($i = 1,2,3$). (b) $X = \{x\}$, $x \mapsto 1,2,3$ in $\mathbf{Q}$, (c) $k\langle X\rangle_i$, $i=3,4,5$.

2. 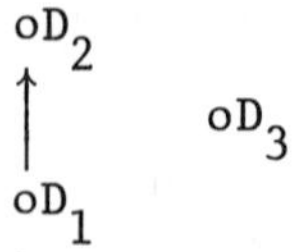 $\circ D_2$ ↑ $\circ D_1$, $\quad \circ D_3$ — Here D_1 specializes to D_2, $D_1 \wedge D_2 \wedge D_3 = (D_1 \wedge D_2) \times D_3$ (a) $X = \{x,y,z\}$, $D_1 = \mathbf{Q}(x,y)$ $(z \mapsto 0)$, $D_2 = \mathbf{Q}(x)$ $(y,z \mapsto 0)$, $D_3 = \mathbf{Q}(z)(x,y \mapsto 0)$ (b) $D_1 = k\langle X\rangle_4$, $D_2 = k\langle X\rangle_2$, $D_3 = k\langle X\rangle_3$.

3. 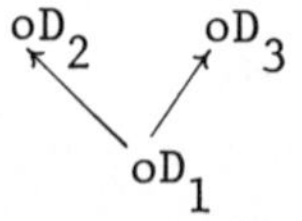 $\circ D_2 \nwarrow \circ D_1 \nearrow \circ D_3$ — $R = D_1 \wedge D_2 \wedge D_3$ semilocal domain $\subseteq (D_1 \wedge D_2) \cap (D_1 \wedge D_3) \subseteq D_1$

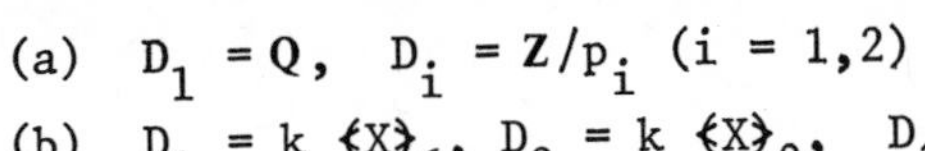

(a) $D_1 = \mathbf{Q}$, $D_i = \mathbf{Z}/p_i$ $(i = 1,2)$

(b) $D_1 = k \langle X \rangle_6$, $D_2 = k \langle X \rangle_2$, $D_3 = k\langle X \rangle_3$.

4. $R = D_1 \wedge D_2 \wedge D_3$ is local ring with two minimal prime ideals, subdirect product of $D_1 \wedge D_3$ and $D_2 \wedge D_3$.

(a) $D_1 = \mathbf{Q}(x)$ $(y \longmapsto 0)$, $D_2 = \mathbf{Q}(y)$ $(x \longmapsto 0)$, $D_3 = \mathbf{Q}(x,y \longmapsto 0)$.

(b) $D_1 = \mathbf{Q}(x \longmapsto 0)$, $D_2 = \mathbf{Q}(x \longmapsto p$ prime$)$, $D_3 = \mathbf{Z}/p$ $(x \longmapsto 0)$.

5. $R = D_1 \wedge D_2 \wedge D_3 = D_1 \wedge D_3$.

(a) $D_1 = \mathbf{Q}(x,y)$, $D_2 = \mathbf{Q}(x)$ $(y \longmapsto 0)$, $D_3 = \mathbf{Q}(x,y \longmapsto 0)$. (b) $D_1 = \mathbf{Q}(x)$, $D_2 = \mathbf{Q}(x \longmapsto 0)$, $D_3 = \mathbf{Z}/p$.

In each case there are commutative examples. For the remaining support systems (non-orders) we have of course only non-commutative examples. In each case $s \propto T$ is indicated by drawing an arrow from s to a balloon enclosing T. We also indicate the partially ordered set of primes $\mathfrak{P}_i = \ker(\Lambda D_j \longrightarrow D_i)$; in each case the lowest prime is 0.

6.

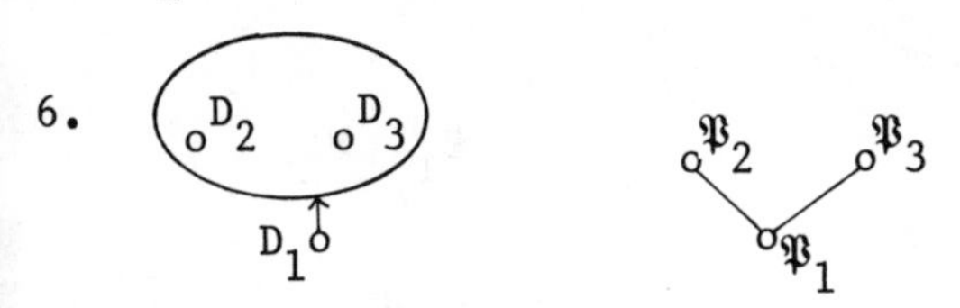

$D_1 = k \langle X \rangle_5$, $D_2 = k \langle X \rangle_2$, $D_3 = k \langle X \rangle_3$.

7.

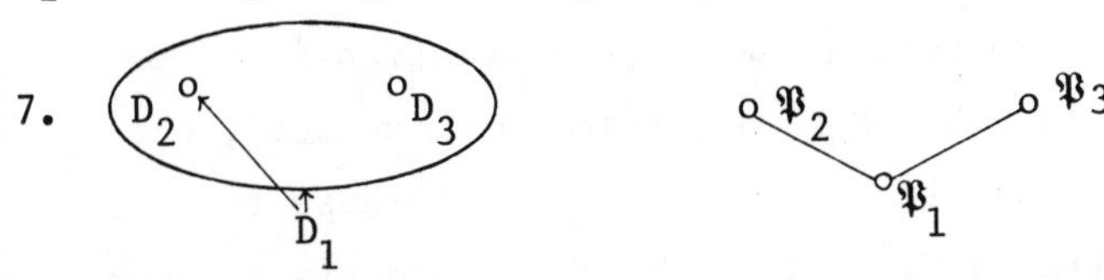

$D_1 = k \langle X \rangle_8$, $D_2 = k \langle X \rangle_2$, $D_3 = k \langle X \rangle_3$.

8.

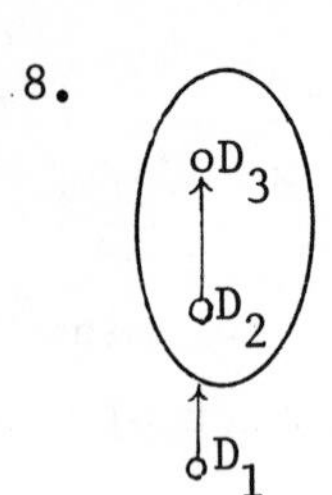

Here $D_1 \propto \{D_3\}$ follows from the relations shown.

$D_1 = k\langle X\rangle_3,\quad D_2 = k\langle X\rangle_2,\quad D_3 = k\langle X\rangle_1.$

9.

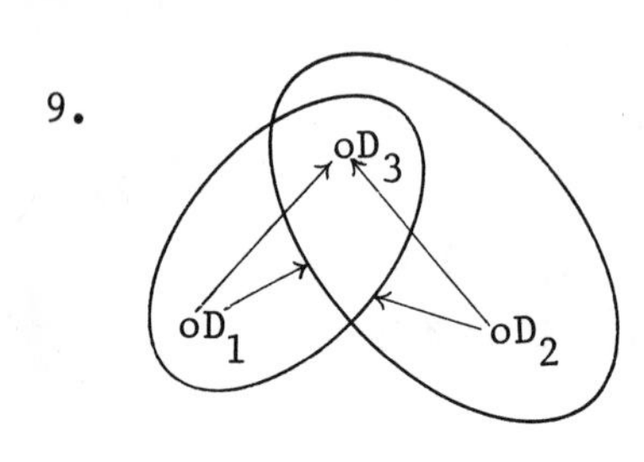

$\mathfrak{P}_3$

$\mathfrak{P}_1 = \mathfrak{P}_2$

$D_1 = k\langle x,y \mid (x^{-1}y)^2 = yx^{-1}\rangle,\quad D_2 = k\langle x,y \mid (x^{-1}y)^3 = yx^{-1}$

$D_3 = k.$

10.

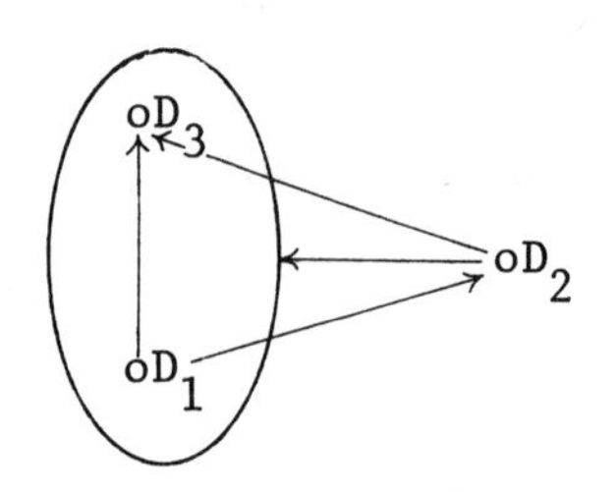

$\mathfrak{P}_3$

$\mathfrak{P}_1 = \mathfrak{P}_2$

$D_1 = k\langle x,y\rangle,\quad D_2 = k\langle x,y \mid (x^{-1}y)^2 = yx^{-1}\rangle,\quad D_3 = k.$

We conclude by expressing the support relation in terms of singular kernels. Let P be any prime matrix ideal, then $\dot{P}$ is defined as the set of matrices all of whose first order minors lie in P. Clearly $\dot{P} \subseteq P$ and under a homomorphism into a field, if P represents the singular matrices, $\dot{P}$ represents the matrices of nullity at least two.

Lemma 7.7.1. *Given any prime matrix ideals* $\mathcal{P}_1,\ldots,\mathcal{P}_r$ *in* k<X>, *any* $n \times n$ *matrix* $A \notin \cup \dot{\mathcal{P}}_i$ *can be extended to an* $n \times n+1$ *matrix which has rank* n mod $\mathcal{P}_i$ *for each* i.

Proof. Write $A = (a_1,\ldots,a_n)$; if a is another column, we put $A^* = (a,a_1,\ldots,a_n)$ and we write $A^* \notin \mathcal{P}_i$ to indicate that A^* has rank n mod $\mathcal{P}_i$; this means that the square matrix obtained by omitting some column of A^* is not in $\mathcal{P}_i$. We use induction on r; when $r = 1$, A has nullity 1 and we can make it non-singular by modifying a single column. When $r > 1$, we can by induction hypothesis adjoin a column to A to obtain an $n \times (n+1)$ matrix A_i such that $A_i \notin \mathcal{P}_j$ $(j \neq i)$. If for some i, $A_i \notin \mathcal{P}_i$, this will show that $A_1 \notin \cup \mathcal{P}_j$; otherwise $A_i \in \mathcal{P}_i$. Now form

$$A_1 \begin{pmatrix} \alpha & 0 \\ 0 & 1 \end{pmatrix} \nabla A_2 \qquad\qquad (\alpha \in k\langle X\rangle).$$

For $\alpha \neq 0$ this is not in $\mathcal{P}_1$ or $\mathcal{P}_2$ and it lies in any $\mathcal{P}_i$ $(i > 2)$ for just one value of α. Avoiding these values (which we can do, because k<X> is infinite) we get a matrix A^* such that $A^* \notin \cup \mathcal{P}_i$. ■

Now the support relation is described by

Theorem 7.7.2. *Let* $\{D_s\}_S$ *be a family of epic* X*-fields and* $\mathcal{P}_s$ *the singular kernel of* D_s, *then* D_t *supports* D_s $(s \in S)$ *if and only if*

$$(1) \quad \mathcal{P}_t \subseteq (\cap \mathcal{P}_s) \cup \cup \dot{\mathcal{P}}_s.$$

In words: *every matrix which becomes singular in* D_t *is either singular in each* D_s *or of nullity* > 1 *in some* D_s.

Proof. Suppose that t supports S and let $A \in \mathcal{P}_t$, $A \notin \cup \dot{\mathcal{P}}_s$. By the lemma we can find a column a such that $(a,A) = A^* \notin \cup \mathcal{P}_s$. Hence the equation $A^*u = 0$ defines $u = (u_o,u_1,\ldots u_n)$ up to a scalar multiple in any D_s.

Since $A \in \mathcal{P}_t$, $u_o = 0$ in D_t and so $u_o = 0$ in all D_s, i.e.

$A \in P_s$ for all s, hence (1) holds.

Conversely, assume (1) and let f be defined in all D_s and f = 0 in D_t. We can find a denominator for f, say A, with numerator A_1, then $A_1 \in P_t$, so either $A_1 \in \cap P_s$, i.e. f = 0 in all D_s, or $A_1 \in \dot{P}_s$ for some s. But then $A \in P_s$ and this contradicts the fact that A was a denominator. ■

8·Equations and singularities

8.1 Equations over skew fields

In the commutative case there is a well known theorem, going back to Kronecker, which asserts that every polynomial equation of positive degree over a commutative field k has a solution in some extension field of k. One effect of this result has been to try to reduce any search for solutions to a single equation. E.g. to find the eigenvalues of a matrix A we solve the equation $\det(xI-A) = 0$.

In the general case no such simple theorem exists (so far!) and in any case we do not have a good determinant function (the determinant introduced by Dieudonné [43] is not really a polynomial but a rational function), so the above reduction is not open to us. In fact we shall find it more profitable to go from scalar equations to matrices.

Our first problem is to write down the general equation in one variable x over a skew field K. We cannot allow x to be central if we want to be able to substitute non-central values of K, but some elements of K are bound to commute with x, e.g. 1, -1 etc. and it is clear that these elements form a subfield k. Moreover, if $\alpha \in k$, so that $\alpha x = x\alpha$, then α must lie in the centre of K if arbitrary substitutions of x are to be allowed. Thus we have a field K which is a k-algebra, and a polynomial in x is an element p of $F = K_k\langle x\rangle = K \sqcup_k k[x]$. Explicitly p has the form

$$(1) \quad a + b_1xc_1 + \ldots + b_rxc_r + d_1xe_1xf_1 + \ldots + d_sxe_sxf_s + \ldots,$$

where $a, b_i, \ldots, \varepsilon K$. Thus even a polynomial of quite low degree can already have a complicated form, and the problem of finding solutions seems at first sight quite hopeless. (But we note that for polynomials in two variables over k, i.e. elements of $k<x,y>$, an extension field containing solutions has been found, by Makar-Limanov [75,77]). A little light can be shed on the problem by trying to generalize it. Instead of finding extensions L of K where $p = 0$ has a root, let us look for L such that a given matrix A over F becomes singular. We shall need some definitions; let us recall that a K-ring is just a ring R with a homomorphism $K \longrightarrow R$ and a homomorphism $f: R \longrightarrow R'$ between K-rings is a K-ring map if the triangle shown commutes.

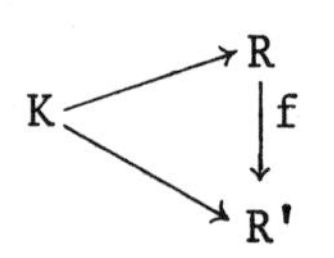

Let A be any square matrix over a K-ring R; we shall say that A is *proper* if there is a K-ring homomorphism of R into a K-field L such that A maps to a singular matrix over L; otherwise A is said to be *improper*. To elucidate this concept we note

Proposition 8.1.1. *Let* K *be any field and* R *a* K-*ring, then an invertible matrix over* R *is improper. When* R *is commutative, the converse holds: every improper matrix is invertible, but this does not hold generally.*

Proof. If A is invertible over R and $f: R \longrightarrow L$ is a K-ring map then A^f is again invertible, hence A is then improper. When R is commutative and A is not invertible, then det A is also a non-unit and so is contained in a maximal ideal $\mathfrak{m}$ of R. The natural map $R \longrightarrow R/\mathfrak{m}$ is clearly a K-ring map into a field and it maps det A to 0, so A becomes singular over $R/\mathfrak{m}$.

To find a counter-example we can limit ourselves to 1×1 matrices, thus we must find a non-unit of R which maps to a unit under any homomorphism into a field. Take a ring with no R-fields, e.g. K_2 ; any element c of K_2 which is not 0 or a unit is improper but not invertible. ∎

Our original problem was this: Does every polynomial p in $F = K_k\langle x\rangle$ which is a nonunit have a zero in some K-field L? We note that p has the zero α in L if and only if the K-ring map $F \longrightarrow L$ defined by $x \mapsto \alpha$ maps p to zero. Thus the question now becomes: Is every non-invertible element of F proper? and it is natural to subsume this under the more general form:

Problem 1. *Is every non-invertible matrix over* $K_k\langle x\rangle$ *proper?*

We shall find that this is certainly not true without restriction on k and K, and find a positive answer in certain special cases, but the general problem is still open. As a first condition we have the following result (Cohn [76']).

Theorem 8.1.2. *Let* K *be a field with centre* k. *If every non-constant equation over* K *has a solution in some extension, then* k *is algebraically closed in* K.

Proof (*) Let $a \in K$ be algebraic over k but not in k, then

$$(2) \qquad ax - xa = 1 \qquad\qquad \text{(metro-equation)}$$

has a solution. Let f be the minimal polynomial for a over k, then by (2), $0 = f(a)x-xf(a) = f'(a)$, where f' is the formal derivative of f. By the minimality of f it follows that $f' = 0$, so a is not separable over k. In particular, this shows that k must be separably closed in K. If a is purely inseparable over k, say $a^q \in k$, where $q = p^r$, then on writing D_a for the mapping $x \mapsto ax - xa$, we have $D_a^q = D_{a^q} = 0$. Since $a \notin k$, $D_a \neq 0$, say $bD_a \neq 0$ for some $b \in K$. Now the equation $xD_a^{q-1} = b$ has a solution x_o in some extension,

(*) I am indebted to P.v.Praag for simplifying my original proof.

but then $x_o D_a^q = bD_a \neq 0$, a contradiction. Hence k must be algebraically closed in K. ■

Thus for Problem 1 to have a positive solution it is necessary for k to be algebraically closed in K. To simplify Problem 1 we introduce a relation between matrices. Let R be any ring, then two matrices A, B over R are said to be *associated* if there are invertible matrices U,V such that A = UBV. For simplicity we assume that R has invariant basis number, so that all invertible matrices are square. Then it is clear that associated matrices have to be the same size, not necessarily square. The following weaker equivalence relation is often useful: Two matrices A,B are said to be *stably associated* if for some unit matrices, $A \dotplus I$ is associated to $B \dotplus I$. Clearly this is again an equivalence relation (for the interpretation in terms of modules see Cohn [76]); stably associated matrices need not be the same size, but if an m x n matrix is stably associated to an r x s matrix then m-n = r-s (at least when R has invariant basis number). The following result is an immediate consequence of the definitions.

Proposition 8.1.3. *Let* R *be a* K-*ring; if a matrix* A *is invertible, or proper, then any matrix stably associated to* A *has the same property.* ■

We can now view the 'linearization by enlargement', already encountered in 6.4, in a different light: It tells us that every matrix over $K_k\langle x\rangle$ is stably associated to a matrix linear in x:

Theorem 8.1.4. *Let* A = A(x) *be a square matrix over* $K_k\langle x\rangle$, *then* A *is stably associated to* Bx + C, *where* $B,C \in K_n$, *for some* n. *Moreover, if* A(0) *is non-singular, we can take* C = I.

Proof. The first part follows as in 6.4. If A(0) is non-singular, then so is C, the result of putting x = 0 in Bx + C, and Bx + C is associated to $C^{-1}Bx + I$. ■

The linear matrix Bx + C obtained here is called the

companion matrix for A(x). To give an example, let

(3) $p(x) = x^n + a_1x^{n-1} + \dots + a_n,$

and write $p = a_n + p_1x$, then the first step is

$$p \longrightarrow \begin{pmatrix} x & -1 \\ a_n & p_1 \end{pmatrix}$$

where we have interchanged the rows. If we continue in this way, we obtain the matrix

$$\begin{pmatrix} x & -1 & 0 & 0 & \dots & 0 \\ 0 & x & -1 & 0 & \dots & 0 \\ & \dots & \dots & & \dots & \\ 0 & 0 & \dots & & \dots & -1 \\ a_n & a_{n-1} & \dots & \dots & & x+a_1 \end{pmatrix}$$

This is of the form $xI - A$, where $A \in K_n$, and this matrix or also A itself is usually known as the companion matrix of p. As we have seen, the general polynomial is more complicated than p and its companion matrix, which always exists, by Th.8.1.4, need not be of the form $xI - A$. Moreover, the companion matrix of a given polynomial or matrix is not generally unique, but it can be shown that if $xI - A$, $xI - B$ are companions for the same matrix then A and B are similar over k (cf. Cohn [76]).

An element p of $K_k\langle x\rangle$ or more generally a matrix P whose companion matrix can be put in the form $xI - A$ is said to be *non-singular at infinity*. E.g. (3) is non-singular at infinity. Generally a matrix P with companion matrix $Bx + C$ is non-singular at infinity if B is non-singular; in any case the rank of B depends only on P and not on the choice of the companion matrix, as is easily seen. This rank is also called the *degree* of P. For a polynomial of the form (2) it clearly reduces to the usual degree.

Let $A \in \mathfrak{M}_n(K)$, by a *singular eigenvalue* of A we understand an element $\alpha \in K$ such that $A - \alpha I$ is singular. We can now state two conjectures whose proof would entail a positive solution of Problem 1.

Conjecture 1. *Every square matrix over a field* K *has a singular eigenvalue in some extension field of* K.

Conjecture 2. *Let* K *be a field which is a* k-*algebra and assume that* k *is algebraically closed in* K. *Then every square matrix* A *over* K *has a non-zero singular eigenvalue in some extension of* K *unless* A *is triangularizable over* k.

If the answer were known, we could settle Problem 1 as follows. Let A be a non-invertible square matrix over $K_k\langle x\rangle$; we have to show that A is proper, and by Th.8.1.4 and Prop.8.1.3 we can instead of A take its companion matrix $Bx + C$. If C is singular, this must be proper, by putting $x = 0$; otherwise we can take it in the form $I - Bx$. If B has a non-zero singular eigenvalue β say, then $I - B\beta^{-1}$ is singular; otherwise by Conjecture 2, there exists $P \in GL_n(k)$ such that $P^{-1}BP = T$ is triangular, hence $P^{-1}(I - Bx)P = I - Tx$ (here we have used the fact that the entries of P lie in k). If T has a non-zero diagonal element α, then $I - T\alpha^{-1}$ is singular; hence all diagonal elements of T are 0 and so $(Tx)^n = 0$; it follows that $I - Tx$ is invertible, hence so is A, a contradiction.

Before we can describe the special cases in which these conjectures can be settled we need to introduce another kind of eigenvalue, which always exists and which can be used to accomplish a transformation to Jordan canonical form.

8.2 Left and right eigenvalues of a matrix

One of the main uses of eigenvalues in the commutative case is to effect a reduction to diagonal form (when possible). Let A be a square matrix over a field K and suppose that A is similar to a diagonal matrix $D = \text{diag}(\alpha_1,\ldots,\alpha_n)$. Then there is a non-singular matrix U such that

$$AU = UD.$$

If we denote the columns of U by $u_1,\ldots,u_n$, this equation can also be written as

$$Au_i = u_i\alpha_i \qquad (i = 1,\ldots,n).$$

This makes it clear that we have indeed an eigenvalue problem, but the α_i need not be singular eigenvalues of A, since they do not in general commute with the components of u_i.

Let K be any field and $A \in K_n$; an element $\alpha \in K$ is called a *right eigenvalue* of A if there is a non-zero column vector u, called an *eigenvector* for α such that

$$(1) \qquad Au = u\alpha.$$

Similarly a *left eigenvalue* of A is an element $\beta \in K$ for which there exists a non-zero row vector v, an *eigenvector* for β, such that $vA = \beta v$. The set of all left and right eigenvalues of A is called the *spectrum* of A, spec A.

Let $c \in K^*$; if $Au = u\alpha$, then $A.uc = u\alpha c = uc.c^{-1}\alpha c$. This shows that the right eigenvalues of A consist of complete conjugacy classes; similarly for left eigenvalues. If $P \in GL_n(K)$, then $P^{-1}AP.P^{-1}u = P^{-1}Au = P^{-1}u\alpha$, hence α is also a right eigenvalue of $P^{-1}AP$. In other words, right (and left) eigenvalues are similarity invariants of A. For singular eigenvalues this is not in general the case; in fact it is easy to see that the three notions of eigenvalue coincide for elements in the centre of K ("central" eigenvalues), but in general there is no very close relation between them. Thus it is possible for a matrix to have a right but no left eigenvalue (Cohn [73"]), but as we shall see later, over an existentially closed field the notions of left and right eigenvalue coincide.

G.M. Bergman has observed (in correspondence) that left and right eigenvalues are special cases of the following more general notion: If $A \in K_n$ then $\alpha \in K$ is called an *inner eigenvalue*, more precisely an r-*eigenvalue* (where $1 \leq r \leq n$) if K^n regarded as a space on which A acts, has an A-invariant subspace W of dimension r-1 and there is $u \notin W$ such that $Au \equiv u\alpha \pmod W$. Then right eigenvalues are just 1-eigenvalues and left eigenvalues are n-eigenvalues (cf. also Sizer [77]).

In order to achieve the transformation to diagonal form one needs a basis of eigenvectors and here one usually applies the well known result that eigenvectors for different eigenvalues are linearly independent. Over a skew field this takes the following form:

Proposition 8.2.1. *Let* A *be a matrix over a field* K, *then the eigenvectors belonging to inconjugate right eigenvalues are linearly independent. If* α *is a right and* β *a left eigenvalue of* A *and* α,β *are not conjugate, then the eigenvectors belonging to them are orthogonal, i.e. if* $Au = u\alpha$, $vA = \beta v$, *then* $vu = 0$.

Proof. Let $\alpha_1,\ldots,\alpha_r$ be right eigenvalues and $u_1,\ldots,u_r$ corresponding eigenvectors and assume that the u's are linearly dependent. By taking a minimal linearly dependent set we may assume that

$$u_1 = \Sigma_2^r u_i\lambda_i \qquad (\lambda_i \in K).$$

By definition $u_1 \neq 0$, hence by minimality $\lambda_i \neq 0$ and $r > 1$. Now $u_1\alpha_1 = Au_1 = \Sigma Au_i\lambda_i = \Sigma u_i\alpha_i\lambda_i$. Hence $\Sigma_2^r u_i(\lambda_i\alpha_1 - \alpha_i\lambda_i) = 0$, but $u_2,\ldots,u_r$ are linearly independent, so $\alpha_i = \lambda_i\alpha_1\lambda_i^{-1}$ and the α_i are all conjugate.

Next if $Au = u\alpha$, $vA = \beta v$, then $vAu = vu.\alpha = \beta.vu$, and if $vu \neq 0$, this would mean that α, β are conjugate. ■

As a corollary we obtain an analogue of a theorem of Gordon and Motzkin [65] which states that the zeros of an equation of degree n cannot lie in more than n conjugacy classes:

Corollary. Spec A *consists of at most* n *conjugacy classes, where* n *is the order of* A. ■

To describe the conditions for diagonalisability we need

Lemma 8.2.2. *Let* R, S *be* k-*algebras and* M *an* (R,S)-*bimodule. Given* $a \in R$, $b \in S$, *assume that there is a polynomial* f *over* k *such that* f(a) *is a unit while* f(b) = 0. *Then for any* $m \in M$ *the equation*

$$(2) \quad ax - xb = m$$

has a unique solution $x \in M$.

Proof. If in $\mathrm{End}_k(M)$ we define $\lambda_a : x \longmapsto ax$, $\rho_b : x \longmapsto xb$, then (2) may be written

$$(3) \quad x(\lambda_a - \rho_b) = m.$$

We note that $\lambda_a\rho_b = \rho_b\lambda_a$ and by hypothesis $f(\lambda_a)$ is a unit and $f(\rho_b) = 0$. Now define the polynomial $\phi(s,t)$ in the commuting variables s,t by

$$\phi(s,t) = \frac{f(s) - f(t)}{s - t},$$

then

$$\phi(\lambda_a,\rho_b)(\lambda_a - \rho_b) = (\lambda_a - \rho_b)\phi(\lambda_a,\rho_b) = f(\lambda_a) - f(\rho_b) = f(\lambda_a).$$

Since $f(\lambda_a)$ is a unit, $\lambda_a - \rho_b$ has a two-sided inverse and it follows that (3) has a unique solution in M. ■

The significance of the lemma lies in this: Given R,S,M as in the lemma, the set of all matrices $\begin{pmatrix} r & m \\ 0 & s \end{pmatrix}$, $r \in R$, $s \in S$, $m \in M$, is a ring under the usual matrix multiplication, and (2) shows that

$$\begin{pmatrix} 1 & x \\ 0 & 1 \end{pmatrix} \begin{pmatrix} a & m \\ 0 & b \end{pmatrix} = \begin{pmatrix} a & 0 \\ 0 & b \end{pmatrix} \begin{pmatrix} 1 & x \\ 0 & 1 \end{pmatrix}$$

i.e. $\begin{pmatrix} a & m \\ 0 & b \end{pmatrix}$ is similar to a 'diagonal' matrix.

Theorem 8.2.3. *Let* K *be any field and* $A \in K_n$. *Then spec* A *cannot contain more than* n *conjugacy classes, and when it consists of exactly* n *classes, all except at most one algebraic over the centre of* K, *then* A *is similar to a diagonal matrix.*

Proof. We have seen that spec A consists of conjugacy classes. Let r be the number of classes containing right eigenvalues and s the number of the remaining classes in spec A, then the space spanned by the columns corresponding to right eigenvalues is at least r-dimensional and the space of rows orthogonal to this is at least s-dimensional, hence $r+s \leq n$, and r+s is just the number of conjugacy classes in spec A.

Assume now that r+s = n; let $\alpha_1,\ldots,\alpha_r$ be inconjugate right eigenvalues and $u_1,\ldots,u_r$ the corresponding eigenvectors, while $\beta_1,\ldots,\beta_s$ are the left eigenvalues not conjugate among themselves or to the α's, with corresponding eigenvectors $v_1,\ldots,v_s$. By Prop.8.2.1 the u's are right linearly independent, the v's are left linearly independent and $v_j u_i = 0$ for all i,j. Write U_1 for the n x r matrix consisting of the columns $u_1,\ldots,u_r$ and V_2 for the s x n matrix consisting of the rows $v_1,\ldots,v_s$. Since the columns of U_1 are linearly independent, we can find an r x n matrix

V_1 over K such that $V_1U_1 = I$ and similarly there is an $n \times s$ matrix U_2 such that $V_2U_2 = I$. Put $U = (U_1 \ U_2)$, $V = \begin{pmatrix} V_1 \\ V_2 \end{pmatrix}$, then

$$VU = \begin{pmatrix} V_1U_1 & V_1U_2 \\ V_2U_1 & V_2U_2 \end{pmatrix} = \begin{pmatrix} I & W \\ 0 & I \end{pmatrix}.$$

The matrix on the right is clearly invertible and since one-sided inverses over a field are two-sided (i.e. a field is weakly finite, cf. Cohn [71"]), we have $U(VU)^{-1} = V^{-1}$, so

$$AV^{-1} = A(U_1 \quad U_2 - U_1W) = (u_1\alpha_1, \ldots, u_r\alpha_r, A(U_2 - U_1W))$$

$$VA = \begin{pmatrix} V_1A \\ \beta_1v_1 \\ \vdots \\ \beta_sv_s \end{pmatrix}.$$

It follows that $VAV^{-1} = \begin{pmatrix} \alpha & T \\ 0 & \beta \end{pmatrix}$, where $\alpha = \operatorname{diag}(\alpha_1, \ldots, \alpha_r)$, $\beta = \operatorname{diag}(\beta_1, \ldots, \beta_s)$ and $T \in {}^rK^s$. Now all the α's and β's are inconjugate and all but at most one are algebraic over the centre of K, hence their minimal equations are distinct (cf.3.3). If only right or only left eigenvalues occur, we have diagonal form; otherwise let $\beta_1, \ldots, \beta_s$ be algebraic, say. Taking f to be the product of their minimal polynomials we have $f(\beta) = 0$ while $f(\alpha)$ is a unit. By Lemma 8.2.2 we can find $X \in {}^rK^s$ such that $\alpha X - X\beta = T$ and transforming our matrix by $\begin{pmatrix} I & X \\ 0 & I \end{pmatrix}$ we reach diagonal form. ■

The restriction on the eigenvalues, that there is to be only one transcendental conjugacy class (at most) is not as severe as appears at first sight, but is to be expected, since K can be extended so that all transcendental elements are conjugate (cf. 5.5).

8.3 Canonical forms for a single matrix over a skew field

As before let K be a field which is a k-algebra; our task is to find a canonical form for a matrix under similarity transformation. The results are not quite as precise as in the commutative case, but come very close, the main difficulty being the classification of polynomials over K, i.e. elements of $K[x]$.

Let $A \in K_n$; in 5.5 we called A *transcendental* over k if for any $f \in k[x]^*$, $f(A)$ is non-singular. If there is a non-zero polynomial f over k such that $f(A) = 0$, A is said to be *algebraic* over k. Of course when $K = k$ (the classical case) every matrix is algebraic, by the Cayley-Hamilton theorem.

In general a matrix is neither algebraic nor transcendental, e.g. $\text{diag}(\alpha, 1)$, where α is transcendental over k, but we have the following reduction.

Proposition 8.3.1. *Every matrix* A *over* K *is similar to the diagonal sum of an algebraic and a transcendental matrix.*

Proof. We can interpret A as as an endomorphism of K^n; clearly being algebraic or transcendental is a similarity invariant, and so may be regarded as a property of the endomorphism. More precisely, $V = K^n$ can be regarded as a $(K, k[t])$-bimodule, where t is a central indeterminate, with the action $vt = Av$. Then the restriction of A to an A-invariant subspace W of V is algebraic if and only if W is a $k[t]$-torsion module. Let V_o be the torsion submodule of V, then V_o is a K-subspace, hence we can find a complement of V_o in V:

$$(1) \qquad V = V_o \oplus V_1.$$

Now A restricted to V_o is algebraic, while the transformation induced on $V_1 \cong V/V_o$ is transcendental. Hence if we use a basis adapted to the decomposition (1), A takes the form

$$\begin{pmatrix} A_o & A' \\ 0 & A_1 \end{pmatrix}$$

where A_o is algebraic and A_1 is transcendental. By applying Lemma 8.2.2 we can reduce A' to 0 and so obtain the desired decomposition. ■

It is not hard to see that the algebraic and transcendental parts are in fact unique up to similarity, so that by this result, we need only consider algebraic or transcendental matrices.

The transcendental part is in many ways simpler to deal with. For by suitably extending K, we can always transform a transcendental matrix to scalar form, as we saw in 5.5. Of course over K itself we cannot expect such a good normal form. We state the result as

Proposition 8.3.2. *Let* K *be existentially closed (over* k*). Then any transcendental matrix* A *is similar to* αI*, where* α *is any transcendental element of* K. ■

To describe the algebraic part, let $V = K^n$ as right K-space, with an algebraic endomorphism Θ. Writing $R = K[t]$ with a central indeterminate t, consider V as right R-module by letting $\Sigma t^i c_i$ correspond to $\Sigma \Theta^i c_i$. If A is the matrix of Θ relative to a basis of V, we shall call V the R-module associated to A. It is clear that two matrices are similar if and only if the associated R-modules are isomorphic.

Now R is a principal ideal domain and every matrix over R is associated to a diagonal matrix (cf. e.g. Cohn [71"], Ch.8), thus there exist $P, Q \in GL_n(R)$ such that

$$(2) \quad P(tI - A)Q = \text{diag}(\lambda_1, \ldots, \lambda_n),$$

where λ_{i-1} is a total divisor of λ_i $(i = 2, \ldots, n)$. This means that for each i there is an invariant element c of R (i.e. $cR = Rc$) such that $\lambda_{i-1}R \supseteq cR \supseteq \lambda_i R$. The λ_i are just the *invariant factors* of $tI - A$ and as right R-module V is isomorphic to the direct sum

$$(3) \quad R/\lambda_1 R \oplus \ldots \oplus R/\lambda_n R.$$

We observe that this holds for any matrix A, algebraic or not. In fact we now see that A is algebraic if and only if λ_n divides a polynomial with coefficients in k. Let us take k to be the precise centre of K, then a polynomial over K is invariant if and only if it is associated to a polynomial over k (Cohn [71"], p.297). It follows that A is algebraic if and only if λ_n divides an invariant polynomial, i.e. (by definition) if and only if λ_n is bounded.

To find when A is transcendental we recall that an element of R is said to be *totally unbounded* if it has no bounded factor (apart from units). Suppose that λ_n has a bounded factor p say, then the R-module V has an element annihilated by p and hence by p^*, where p^* is the bound of p (i.e. the least invariant element divisible by p). Now $p^* = p^*(t)$ is invariant and $p^*(A)$ is singular, so A cannot be transcendenta Conversely, if A is not transcendental, V has an element anni hilated by an invariant polynomial, so some invariant factor λ_i has a factor which is bounded, and hence λ_n then has a bounded factor. This proves most of

Proposition 8.3.3. *Let* K *be a field with centre* k *and let* $A \in K_n$ *have invariant factors* $\lambda_1,\ldots,\lambda_n$. *Then* (i) A *is algebraic over* k *if and only if* λ_n *is bounded,* (ii) A *is transcendental over* k *if and only if* λ_n *is totally unbounded, and then* $\lambda_1 = \ldots = \lambda_{n-1} = 1$.

Only the last part still needs proof: Each λ_i $(i < n)$ is a total divisor of λ_n, so there is an invariant element c such that $\lambda_i \mid c \mid \lambda_n$. But the only invariant element dividing λ_n is 1, hence $\lambda_i = 1$ $(i = 1,\ldots,n-1)$.■

To obtain a normal form for algebraic matrices we need a result on the decomposition of cyclic modules over a principal ideal domain R. We recall (cf. Cohn [71"] p.229) that a cyclic R-module R/aR has a direct decomposition

$$R/aR \cong R/q_1R \oplus \ldots \oplus R/q_\nu R \oplus R/uR,$$

where each q_i is a product of pairwise stably associated bounded atoms, while atoms in different q's are not stably associated, and u is totally unbounded. If we apply this result to (3) and observe that λ_i for $i < n$ is necessarily bounded, we obtain a direct decomposition

$$(4) \qquad V = R/\alpha_1R \oplus \ldots \oplus R/\alpha_rR \oplus R/uR,$$

where each α_i is a product of stably associated bounded atoms (but now atoms in different α's may be stably associated), and u is totally unbounded. The term R/uR corresponds to the transcendental part of Θ and is left unchanged. The R/α_jR correspond to the algebraic parts of Θ; unlike the commutative case these terms need not be indecomposable, but by decomposing them further we may assume each term in (4) indecomposable. The decomposition is then unique, by the Krull-Schmidt theorem (because the algebraic part is itself unique), and the resulting polynomials $\alpha_1,\ldots,\alpha_r$ are the *elementary divisors* of A. They are unique up to stable association. Thus we can write A as a diagonal sum of terms corresponding to the different elementary divisors.

We have already dealt with the transcendental part; let now A be an algebraic matrix with a single elementary divisor α, then α is a product of stably associated atoms, say

$$(5) \qquad \alpha = p_1\ldots p_s.$$

Each p_j has the same degree d say (as polynomial in t) and $sd = n$ is the order of A. Let V be the R-module associated to A, then V is cyclic (being isomorphic to $R/\alpha R$). Let v be a generator, then $v, v\Theta, \ldots, v\Theta^{n-1}$ is a basis of V, for v, as generator of V, cannot be annihilated by a polynomial of degree less than n. We still have a basis if we take $v, v\Theta, \ldots, v\Theta^{d-1}$, $vp_s, v\Theta p_s, \ldots, vp_{s-1}p_s, \ldots, v\Theta^{d-1}p_2 \ldots p_s$. Relative to this basis Θ has the matrix

$$\begin{pmatrix} P_s & N & 0 & 0 & \ldots & 0 \\ 0 & P_{s-1} & N & 0 & \ldots & 0 \\ & \ldots & \ldots & & \ldots & \\ 0 & 0 & 0 & 0 & \ldots P_2 & N \\ 0 & 0 & 0 & 0 & 0 & P_1 \end{pmatrix}$$

where P_i is the companion matrix of p_i and $N = e_{s1}$ is an s x s matrix with 1 in the SW-corner and the rest zeros.

This describes A completely and we obtain an expression much like the rational canonical form in the commutative case. However, unlike the latter, the above expression is not unique; in fact the p_i are determined only up to stable association and not every choice of p's (in the class of stably associated elements) is possible, thus the factors in (5) cannot be prescribed.

There is one case in which we can get a more precise description, namely when every polynomial over k splits into linear factors over K. Even if this does not hold, we can enlarge K to a field L for which it holds by taking $L = K \circ_k \bar{k}$, where $\bar{k}$ is an algebraic closure of k. It now follows that all bounded atoms over K are linear, and for linear polynomials it is easy to describe stable association. We recall from (Cohn [71"], Ch.3) that over a principal ideal domain, a and a' are stably associated (= similar l.c.) if and only if there is a comaximal relation

(6) $ab' = ba'$.

Now assume that a, a' are linear; they may be taken monic, without loss of generality, say $a = t - \alpha$, $a' = t - \alpha'$, then on replacing b in (6) by its remainder after division by a we may assume that $b \in K$, and hence $b' \in K$ (by comparing degrees). If we now compare highest terms, we find that $b' = b$ and hence

(7) $\alpha' = b^{-1}\alpha b$.

Conversely, when (7) holds, $t - \alpha$ and $t - \alpha'$ are stably associated.

To obtain a good analogue of the Jordan normal form we need to assume that k is perfect. The next result and its application is due to W.S. Sizer [75]. A polynomial $f \in R = K[t]$ will be called *indecomposable* if R/fR is indecomposable as right R-module.

Proposition 8.3.4. *Let* K *be a field with centre* k. *Assume that* k *is perfect and that there is a commutative field* F, $k \subseteq F \subseteq K$, *such that every polynomial over* k *splits completely over* F. *Then every indecomposable bounded polynomial over* K *is stably associated to a polynomial of the form* $(t - \alpha)^n$.

Proof. Let $a \in K[t]$ be indecomposable bounded. Its bound has the form p^n (cf. Cohn [71"] p. 231), where p is a monic invariant atom, hence an irreducible polynomial over k. By hypothesis we have

$$p = (t - \alpha_1) \ldots (t - \alpha_r) \qquad \alpha_i \in F,$$

where the α_i are distinct, because k is perfect. We claim that $(t - \alpha_1)^n$ has bound p^n. For clearly $(t - \alpha_1)^n \mid p^n$, hence the bound of $(t - \alpha_1)^n$ is a factor of p^n, say p^m, where $m \leq n$. Thus

$$p^m = (t - \alpha_1)^n q = (t - \alpha_1)^m \prod_{i=2}^{r} (t - \alpha_i)^m.$$

By unique factorization in $F[t]$ we must have $n = m$ because the α_i are distinct, hence p^n is the bound of $(t - \alpha_1)^n$. Clearly $(t - \alpha_1)^n$ is indecomposable and so it is stably associated to a, because it has the same bound (cf. Cohn [71"], p.231).■

Now let K be a field with perfect centre k, and assume that K contains an algebraic closure of k. If A is a matrix over K with a single elementary divisor, then this takes the form $(t - \alpha)^n$. It follows that A is similar to

$$\begin{pmatrix} \alpha & 1 & 0 & 0 & \dots & 0 \\ 0 & \alpha & 1 & 0 & \dots & 0 \\ & \dots & \dots & \dots & & \\ 0 & 0 & \dots & \dots & & \alpha \end{pmatrix}$$

When K has a perfect centre k, we can extend K so as to contain a commutative field F containing an algebraic closur of k as well as an element transcendental over k. By combining the above steps we then can for any matrix over K find a similar matrix over F in Jordan canonical form.

In conclusion we briefly note the case of a field finite-dimensional over its centre. Let K have centre k, then $[K:k] = n^2$ is a perfect square and any maximal commutative subfield of K has degree n. Let F be such a field and $u_1, \dots, u_n$ a left F-basis for K, then for any $a \in K$,

$$(8) \qquad u_i a = \Sigma \rho_{ij}(a) u_j \qquad\qquad \rho_{ij}(a) \in F,$$

and it is easily verified that the mapping $a \longmapsto (\rho_{ij}(a))$ is a k-homomorphism of K into F_n. Since F is commutative, left, right and singular eigenvalues coincide, but of course none need exist in F. Suppose however that $(\rho_{ij}(a))$ has an eigenvalue α in F, then there exist $\gamma_1,\ldots,\gamma_n \in F$, not all 0, such that

$$\Sigma\gamma_i\alpha u_i = \Sigma\gamma_i\rho_{ij}(a)u_j = \Sigma\gamma_i u_i a.$$

Writing $c = \Sigma\gamma_i u_i$, we have $\alpha c = ca$ and $c \neq 0$. Thus $(\rho_{ij}(a))$ has an eigenvalue in F if and only if a is conjugate to an element of F. By the Skolem-Noether theorem this is so whenever k(a) is isomorphic to a subfield of F. All this still applies if a is an r x r matrix over K, if we interpret $(\rho_{ij}(a))$ as an nr x nr matrix over F. Thus a matrix A over K has a left eigenvalue in F if and only if its image under ρ (given by (8)) has an eigenvalue in F. In general this is not much help because F need not be algebraically closed, but in the special case of the quaternions F equals **C**, the complex numbers. Applying the above remarks, we see that every matrix over the quaternions has left and right eigenvalues, and we can make a similarity reduction as indicated above. This is of course well known (cf. Wiegmann [55]); by contrast it is not known (to the author) whether every quaternion matrix has a singular eigenvalue.

8.4 Special cases of the singular eigenvalue problem

In 8.1 we saw that the problem of solving equations can be reduced to the singular eigenvalue problem (conjectures 1,2). Now in 8.3 we have seen how to find left and right eigenvalues of an arbitrary matrix, but this does not on the face of it help us to find singular eigenvalues. Nevertheless there are some special cases which can be solved using the results of 8.3. Firstly there is the case of equations in which all the coefficients are on one side:

Theorem 8.4.1. *Let* K *be a field, then any equation*

$$(1) \qquad x^n + a_1x^{n-1} + \ldots + a_n = 0 \qquad\qquad (a_i \in K)$$

has a solution in some extension of K.

Proof. Consider the companion matrix of this equation:

$$A = \begin{pmatrix} 0 & 1 & 0 & 0 & \ldots & 0 & 0 \\ 0 & 0 & 1 & 0 & \ldots & 0 & 0 \\ & \ldots & & \ldots & & \ldots & \\ 0 & 0 & 0 & 0 & \ldots & 0 & 1 \\ -a_n & -a_{n-1} & \ldots & & \ldots & -a_2 & -a_1 \end{pmatrix}.$$

In a suitable extension we can find a right eigenvalue of A:

$$(2) \qquad Av = v\alpha.$$

In detail this reads: $v_2 = v_1\alpha$, $v_3 = v_2\alpha, \ldots, v_n = v_{n-1}\alpha$, $-a_nv_1 - a_{n-1}v_2 - \ldots - a_1v_n = v_n\alpha$. If $v_1 = 0$, these equations show that $v_2 = \ldots = v_n = 0$, but $v \neq 0$, hence $v_1 \neq 0$ and we may adjust the equation (2) so that $v_1 = 1$. Then $v_2 = \alpha$, $\ldots, v_n = \alpha^{n-1}$ and the last equation reads

$$\alpha^n + a_1\alpha^{n-1} + \ldots + a_n = 0.$$

Thus $x = \alpha$ is a solution of (1). ■

What has happened here is that the eigenvector v has components which are polynomials in α, hence $v\alpha = \alpha v$ and so α is also a singular eigenvalue of A. The above method gives one solution of (1) and as in the commutative case we can use this solution to reduce (1) to an equation with a zero root, but the new equation is no longer of the form (1); neither is it possible in general to reduce the degree.

Our next aim is to solve the singular eigenvalue problem in the special case of 2 x 2 matrices, but for this we need first to consider two special quadratic equations.

Theorem 8.4.2. *Let* K *be any field; given* $a,b,c,d \in K$, $a \neq 0$, *the equation*

$$(3) \quad xax + bx + xc + d = 0$$

has a solution in some extension field of K.

Proof. Let $A = \begin{pmatrix} -b & -d \\ a & c \end{pmatrix}$ and consider the equation $Av = v\lambda$, which we know has a solution (with $v \neq 0$) in some extension of K. If we write $v = (x,y)^T$, this means that we have

$$-bx - dy = x\lambda,$$

$$ax + cy = y\lambda.$$

If $y = 0$, the second equation reduces to $ax = 0$, and so $x = 0$. But x,y cannot both vanish, hence $y \neq 0$ and we can adjust the equations so that $y = 1$. If we now eliminate λ, we get

$$-bx - d = x(ax + c),$$

i.e. (3). ∎

Next we consider (3) when $a = 0$. Changing the notation slightly, we have

$$(4) \quad ax - xb = c.$$

Let $p \in K^*$; if we write $y = xp$, (4) becomes $ayp^{-1} - yp^{-1}b = c$, i.e.

$$ay - y.p^{-1}bp = cp.$$

Thus in (4) we can either replace c by 1 (provided only that $c \neq 0$) or replace b by a conjugate. We also recall from 8.2 that (4) is equivalent to the matrix equation

$$(5) \quad \begin{pmatrix} 1 & x \\ 0 & 1 \end{pmatrix}\begin{pmatrix} a & c \\ 0 & b \end{pmatrix} = \begin{pmatrix} a & 0 \\ 0 & b \end{pmatrix}\begin{pmatrix} 1 & x \\ 0 & 1 \end{pmatrix}$$

Put $A = \begin{pmatrix} a & c \\ 0 & b \end{pmatrix}$, $B = \begin{pmatrix} a & 0 \\ 0 & b \end{pmatrix}$; if a,b are both transcendental over k, the matrices A, B are transcendental over k and by Th. 5.5.5 are similar over an extension of K, i.e. there exists a non-singular matrix P such that PA = BP. Writing $P = \begin{pmatrix} p & q \\ r & s \end{pmatrix}$, we thus have the equations

$$\begin{array}{ll} pa = ap & pc + qb = aq, \\ ra = br & rc + sb = bs. \end{array}$$

Since P is invertible, p,r cannot both vanish, say $p \neq 0$. Then from the first two equations we find that $c + p^{-1}qb = p^{-1}aq = ap^{-1}q$; i.e. $x = p^{-1}q$ is a solution of (4). Similarly if $r \neq 0$, $x = r^{-1}s$ is a solution.

If one of a,b is algebraic over k and the other transcendental, or both are algebraic but with different minimal equations over k, then (4) has a unique solution, by Lemma 8.2.2.

There remains the case where a,b have the same minimal equations over k. Here we need

Lemma 8.4.3. *Let* K *be a field with centre* C *and let* $a,b \in K$ *have the same minimal polynomial* μ *over* C. *Then the equation*

$$(6) \quad ax - xb = 1$$

has a solution in K *(or indeed in any extension field with centre* C*) if and only if in the polynomial ring* $K[t]$, $(t - b)(t - a)$ *divides* $\mu(t)$.

Proof. In the ring $R = K[t]$, $t - a$ and $t - b$ are similar bounded atoms, with the common bound μ. By Prop. 6.5.7 of Cohn [71''] p.231, $(t - b)(t - a) \mid \mu$ if and only if $(t - b)(t - a)$ is decomposable, i.e. there exist $p,q \in R$ such that

$$(t - b)(t - a) = pq, \quad (t - b)R + pR = R = R(t - a) + Rq.$$

By Lemma 3.4.3, l.c. p.130, this is the case if and only if there exist $f,g \in R$ such that

$$(7) \quad f(t - b) - (t - a)g = 1.$$

By the division algorithm in R we can write $f = (t - a) f_1 + u$, where $f_1 \in R$, $u \in K$. Inserting this expression and simplifying, we obtain

$$u(t - b) - (t - a)v = 1,$$

where $v = g - f_1(t - b)$. By comparing terms of highest degree in t we see that $v \in K$ and $v = u$. If we now equate constant terms we get

$$au - ub = 1,$$

and this solves (6). Conversely, when (6) has a solution $x = u$, then (7) holds with $f = g = u$ and so $(t - b)(t - a)$ is decomposable, whence $(t - b)(t - a) \mid \mu$, as claimed. ■

Let us now return to (4); as we have seen, for $c \neq 0$ this is equivalent to $ay - y.cbc^{-1} = 1$. Now if a,b have the same minimal polynomial μ over k, we can enlarge K so that k becomes the exact centre (e.g. by taking the field coproduct of K and k(x) over k). Then (4) has a solution if and only if $(t - cbc^{-1})(t - a) \mid \mu$.

Finally we can ask when the solution is unique; by linearity (4) has a unique solution if and only if $ax = xb$ has no non-zero solution, i.e. a,b are not conjugate. Summing up our results, we obtain

Theorem 8.4.4. Let K *be a field which is a* k*-algebra, and consider the equation*

$$(4) \qquad ax - xb = c \qquad\qquad (a,b,c \in K).$$

(i) *If* a,b *are both transcendental over* k*,* (4) *has infinitely many solutions in a suitable extension of* K*,*

(ii) *If at least one of* a,b *is algebraic over* k *but the other has either a different minimal equation or is transcendental over* k*, then* (4) *has a unique solution in a suitable extension of* K*,*

(iii) *If* a,b *have the same minimal polynomial* μ *over the centre* k *of* K*, then* (4) *has a solution in* K *(or in any extension of* K*) if and only if either* $c = 0$*, or* $(t - cbc^{-1})(t - a)$ *divides* μ*, or equivalently, if* $(t - cbc^{-1})(t - a)$ *is decomposable in* $K[t]$. ■

We can now solve the singular eigenvalue problem for 2 x 2 matrices.

Theorem 8.4.5. Let K *be any field, then every* 2 x 2 *matrix* A *over* K *has a singular eigenvalue in some extension of* K*. If* K *is a* k*-algebra, where* k *is algebraically closed, then* A *has a non-zero singular eigenvalue unless* A *is triangularizable over* k.

Proof. Let $A = \begin{pmatrix} a & b \\ c & d \end{pmatrix}$, $a,b,c,d \in K$. We have to find x such that $A - xI$ is singular. If $c = 0$, we can take $x = a$. Otherwise, on replacing x by cx, we may take $c = 1$. Now

$$\begin{pmatrix} a - x & b \\ 1 & d - x \end{pmatrix} \text{ is associated to } \begin{pmatrix} a - x & b - (a-x)(d-x) \\ 1 & 0 \end{pmatrix}$$

and this is singular precisely if $(a - x)(d - x) = b$, i.e.

$$(8) \qquad x^2 - xd - ax + (ad - b) = 0.$$

By Th. 8.4.2, this has a solution in some extension field.

Suppose now that A has no non-zero singular eigenvalue. If $c = 0$, this means that $a = d = 0$ and A is already upper triangular. Otherwise we may again take $c = 1$ and reach the equation (8). If this has no non-zero solution, we must have $ad = b$, and on writing $y = x^{-1}$ we can bring this to the form

$$dy + ya = 1.$$

By hypothesis this has no solution over any extension of K, hence by Th.8.4.4, d,−a are algebraic over k, with the same minimal equation over k (if they had different minimal equations over k we could reduce the centre to k). But k is algebraically closed in K, hence $a, d \in k$ and $a = -d$. Thus A takes the form

$$A = \begin{pmatrix} a & -a^2 \\ 1 & -a \end{pmatrix} \qquad \text{where } a \in k.$$

But this is clearly similar to $\begin{pmatrix} 0 & 1 \\ 0 & 0 \end{pmatrix}$ over k. ∎

For another case in which the singular eigenvalue problem can be solved see Cohn [76'].

8.5 Specializations and the rational topology

So far we have only been concerned with finding one solution of any equation (where possible), but in algebraic geometry one has to deal with the system of all solutions. Now we can define varieties as solution sets of systems of equations, but very little is known so far about such varieties. Below we shall examine a rather plausible definition of irreducible variety (as the locus of a 'generic' point) and give some examples. Here is a very simple example which already shows the difference between the commutative and the general case.

Let $c \in K$, then one would expect the equation

$$(1) \quad cx - xc = 0$$

to define a zero-dimensional variety (provided that c is not in the centre of K). But the solution set of (1) is a subfield of K (the centralizer of c in K), whereas in the commutative case 0-dimensional varieties are finite sets of points.

Consider a field K which is a k-algebra and let E be a K-field. We shall consider affine n-space over E, E^n or $\mathbf{A}^n(E)$, whose points are described by n-tuples in E. Given α, $\beta \in E^n$, we write $\alpha \longrightarrow \beta$ and call β a *specialization* of α (over K) if the map $\alpha_i \longmapsto \beta_i$ defines a specialization in the sense of 7.3. In terms of singular kernels, if $\mathcal{P} = \mathrm{Ker}(x \longmapsto \alpha)$, $\mathcal{Q} = \mathrm{Ker}(x \longmapsto \beta)$, this means that $\mathcal{P} \subseteq \mathcal{Q}$, or more concretely, for any matrix $A = A(x)$ over $K_k\langle x\rangle$ ($x = (x_1,\ldots,x_n)$),

$$A(\alpha) \text{ singular} \Rightarrow A(\beta) \text{ singular.}$$

Since every matrix over $K_k\langle x\rangle$ is stably associated to a linear matrix, it is enough to require that this holds for linear matrices. Thus we have

Theorem 8.5.1. *Let* E/K *be any extension and* $\alpha, \beta \in E^n$, *then* $\alpha \longrightarrow \beta$ *over* K *if and only if*

$$A + \Sigma A_i\alpha_i \textit{ singular} \Rightarrow A + \Sigma A_i\beta_i \textit{ singular}$$

for any matrices $A, A_1, \ldots, A_n$ *over* K. ■

We shall call α *free* over K if the map $x \mapsto \alpha$ defines an isomorphism $K_k\langle x\rangle \cong K(\alpha)$. Then we have the

Corollary. Let $\alpha \in E^n$, *then* α *is free over* K *if and only if* $A + \Sigma A_i\alpha_i$ *is singular only when* $A + \Sigma A_i x_i$ *is not full in* $K_k\langle x\rangle$. ■

Th. 8.5.1 also makes it clear how specializations in projective space should be defined. Each point of the projective space $\mathbf{P}^n(E)$ is described by an $(n+1)$-tuple $\xi = (\xi_o, \ldots, \xi_n)$ and ξ, η represent the same point if and only if $\xi_i = \eta_i\lambda$ for some $\lambda \in E^*$. At first sight it is not clear how specialization in projective space is to be defined; instead of polynomials we would have to consider rational functions in the x_i and (in contrast to the commutative case) there is no simple way of getting rid of the denominators. However, with Th. 8.5.1 in mind we can define $\xi \longrightarrow \eta$ (over K) if and only if

$$\Sigma_o^n A_i\xi_i \text{ singular} \Rightarrow \Sigma_o^n A_i\eta_i \text{ singular},$$

for any matrices $A_o, \ldots, A_n$ over K.

The condition of Th. 8.5.1 can still be simplified if we are specializing to a point in K:

Theorem 8.5.2. *Let* E/K *be any extension,* $\alpha \in E^n$, $\lambda \in K^n$, *then* $\alpha \longrightarrow \lambda$ *if and only if for any matrices* $A_1, \ldots, A_n$ *over* K, $I - \Sigma A_i(\alpha_i - \lambda_i)$ *is non-singular.*

Proof. Assume that $\alpha \longrightarrow \lambda$; if $I - \Sigma A_i(x_i - \lambda_i)$ becomes singular for $x = \alpha$, it must also be singular for $x = \lambda$, but then it reduces to I, a contradiction, hence $I - \Sigma A_i(\alpha_i - \lambda_i)$ is non-singular.

Conversely, when the condition is satisfied, let $\mathcal{P}$ be the singular kernel of the map $K_k\langle x\rangle \longrightarrow K(\alpha)$. We must show that under the map $x \mapsto \lambda$ every matrix of $\mathcal{P}$ becomes singular, and here it is enough to test matrices of the form $A + \Sigma A_i x_i$. Thus let $A + \Sigma A_i \alpha_i$ be singular, we have to show that $C = A + \Sigma A_i \lambda_i$ is also singular; note that C has entries in K. If C were non-singular, we could write

$$A + \Sigma A_i \alpha_i = A + \Sigma A_i \lambda_i + \Sigma A_i (\alpha_i - \lambda_i)$$

$$= C + \Sigma A_i (\alpha_i - \lambda_i)$$

$$= C(I - \Sigma B_i (\alpha_i - \lambda_i)), \text{ where } B_i = -C^{-1} A_i.$$

Here the left-hand side is singular, and by hypothesis the right-hand side is non-singular, a contradiction, which shows that $A + \Sigma A_i \lambda_i$ is in fact singular. ■

With the help of the specialization lemma 6.3.1 we again get a criterion for a point to be free:

Corollary. Let K *be infinite-dimensional over its centre* k, *where* k *is infinite, then for any* $\alpha \in E^n$ *the following conditions are equivalent:*

(a) α *is free over* K,

(b) *every point of* K^n *is a specialization of* α,

(c) $I - \Sigma A_i (\alpha_i - \lambda_i)$ *is non-singular for all* A_i *over* K *and all* $\lambda_i \in K$.

Proof. By Th. 8.5.2, (b) $\Longleftrightarrow$ (c) and (a) $\Rightarrow$ (b) is clear. To prove (b) $\Rightarrow$ (a), let $A + \Sigma A_i x_i$ be full, then by the specialization lemma 6.3.1, there exists $\lambda \in K^n$ such that $A + \Sigma A_i \lambda_i$ is non-singular, hence by (b), $A + \Sigma A_i \alpha_i$ is non-singular, and this shows α to be free over K. ■

To restate this Cor. we introduce the following definitions. A point $\alpha \in E^n$ is called an *inverse eigenvalue* of the sequence $A_1, \ldots, A_n$ if $I - \Sigma A_i \alpha_i$ is singular. Secondly, given any extension E/K, we can regard the vector space K^n as a subgroup of E^n; its cosets will be called the *levels* in E over K. Thus $\alpha, \beta \in E^n$ are on the same level precisely if $\alpha - \beta \in K^n$. Now the above Cor. may be stated by saying that α is free over K if and only if its level contains no inverse eigenvalue of any sequence of matrices over K.

Let $K \subseteq E \subseteq L$; given $\alpha \in L^n$, we define the *locus* of α in E over K as the set of all specializations of α in E over K.

Examples 1. If $\alpha \in K^n$, the locus of α in K is just the point α.

2. Let K be infinite-dimensional over an infinite centre. Then α is free over K if and only if the locus of α in K is all of K^n.

3. If $A \in K_r$ has an inverse eigenvalue α in L but none in E, then the locus of α in E is empty.

To illuminate the situation in the general case let us take the case of commutative fields $E \supseteq K$. If $\alpha \in E$ is algebraic over K but not in K, then α satisfies an equation

$$(2) \quad f(x) = 0,$$

over K. If (2) also has a root λ in K, we can replace f by a polynomial of lower degree which still has α as zero but not λ. In the general case it may not be possible to separate out the rational solutions in this way; those that always accompany α represent the locus. To give an example of this behaviour we shall construct a point α which has a specialization in K without itself being in K or free over K.

In the example the locus of α consists of precisely one point, viz. ∞. The condition at ∞ means that $I - A\alpha^{-1}$ is non-singular for all A, i.e. $\alpha I - A$ is non-singular, while the condition at a point $\lambda \in K$ means that $I - A(\alpha - \lambda)$ is singular for some A. Let k be a commutative field of characteristic 0 and form the rational function field $k(t)$;

now take K to be the universal field of fractions of the free algebra $k(t)\langle u_1, u_2\rangle$, then K is infinite-dimensional over its centre k(t) and k is infinite.

Let α be a root of

$$(3) \qquad (x + 1)t - tx = 0$$

in a suitable extension field of K (which exists by Th. 8.4.4). Linearizing the left-hand side, we get

$$I - Cx, \qquad \text{where } C = \begin{pmatrix} 1_{-1} & -t \\ t & -1 \end{pmatrix} .$$

This shows that α cannot be specialized to 0 (this is already clear from (3): putting x = 0 in (3), we get t = 0, a contradiction). Since t is in the centre of K, $\alpha - \lambda$ satisfies (3) for any $\lambda \in K$, hence α cannot be specialized to any $\lambda \in K$. To verify the condition at ∞ we recall that a matrix of order n over a field K cannot have more than n central eigenvalues, by Lemma 6.3.3. Now by (3), $t\alpha = (\alpha + 1)t$ and $At = tA$ for any matrix A over K, hence

$$t(\alpha I - A) = ((\alpha + 1)I - A)t .$$

Hence if $A - \alpha I$ is singular, so is $A - (\alpha + 1)I$, and by induction, $A - (\alpha + \nu)I$, for all $\nu \in \mathbf{Z}$. Since K has characteristic 0, this means that $A - \alpha I$ has infinitely many central eigenvalues, a contradiction. Thus $A - \alpha I$ cannot be singular and it follows that $\alpha \longrightarrow \infty$.

If we extend K to a field E which contains a root s of (3), then the solution α of (3) has infinitely many specializations in E, viz. ∞ and s, and hence the whole level of s over K, but no point of K itself.

We have already met the rational topology in 7.2. It seems natural to define it in terms of matrices rather than rational equations. To do so we define for each matrix $A = A(x)$ over $K_k\langle x\rangle$ a subset of E^n, its *singularity support*, as

$$D(A) = \{\alpha \in E^n \mid A(\alpha) \text{ is non-singular}\}.$$

Now the *rational* K-*topology* on E^n may be defined by taking the $D(A)$ as a base for the open sets. If E satisfies the conditions of the specialization lemma, it is easy to see that $D(A)$ is non-empty precisely when A is full. Moreover E^n is irreducible: if $D(A)$, $D(B)$ are not empty, then A, B are full, hence so is $A \dotplus B$ and $D(A) \cap D(B) = D(A \dotplus B) \neq \emptyset$. It may happen that all non-empty open sets have a non-empty intersection. By Th. 8.5.1 Cor., $\alpha \in E^n$ is free over K if and only if it lies in every non-empty open subset of E^n. This remark shows that the rational E-topology is generally finer than the rational K-topology.

The singularity support of any matrix is clearly the same as that of its companion matrix, so in discussing supports we need only consider linear matrices. When k is infinite and K infinite-dimensional over its centre it is even enough to take matrices of the form $I - \Sigma B_i(x_i - \lambda_i)$, by the argument used to prove Th. 8.5.2, Cor. In this way we see that for a given $\lambda \in K^n$ the sets $D(I - \Sigma B_i(x_i - \lambda_i))$ form a neighbourhood base for λ. Now we can identify the locus of a point as its closure in the rational topology:

Theorem 8.5.3. *Let* K *be a* k-*algebra. Given* $E \supseteq K$ *and* $\alpha \in E^n$, *the locus of* α *in* K *is the closure of the point* α *in the rational* K-*topology.*

Proof. Let $\lambda \in K^n$, then $\lambda \in \overline{\{\alpha\}}$ if and only if α lies in every neighbourhood of λ, i.e. $I - \Sigma B_i(\alpha_i - \lambda_i)$ is non-singular for all B_i over K. But by Th. 8.5.2 this is just the condition that $\alpha \longrightarrow \lambda$. ∎

8.6 Algebraic dependence

Let E/K be a skew field extension, where E, K are k-algebras. Given $\alpha_1,\ldots,\alpha_n,\beta \in E$, we say that β is *algebraically dependent* on $\alpha_1,\ldots,\alpha_n$ over K if there is a matrix $A = A(x_1,\ldots,x_n,y)$ over $K_k\langle x_1,\ldots,x_n,y\rangle$ such that $A(\alpha_1,\ldots,\alpha_n,y)$ is full but $A(\alpha_1,\ldots,\alpha_n,\beta)$ is singular. By linearization we may take A to be linear in its arguments, but this is not needed in what follows. It is natural to ask whether the notion of algebraic dependence thus defined satisfies the usual axioms for a dependence relation; this would of course have far-reaching consequences. Unfortunately not all the axioms hold. We briefly recall the definitions (cf. e.g. Cohn [77], Ch.1).

An abstract *dependence relation* on a set S associates with each finite subset X of S certain elements of S, said to be *dependent* on X, subject to the conditions:

D.0 *If* $X = \{x_1,\ldots,x_n\}$, *then each* x_i *is dependent on* X,

D.1 *(transitivity) If* z *is dependent on* $\{y_1,\ldots,y_m\}$ *and each* y_j *is dependent on* $\{x_1,\ldots,x_n\}$, *then* z *is dependent on* $\{x_1,\ldots,x_n\}$,

D.2 *(exchange axiom) If* y *is dependent on* $\{x_1,\ldots,x_n\}$, *but not on* $\{x_2,\ldots,x_n\}$, *then* x_1 *is dependent on* $\{y,x_2,\ldots,x_n\}$.

Linear dependence (in a vector space over a field) and algebraic dependence over a commutative field are familiar examples. We observe that the above notion of algebraic dependence satisfies D.0 and (usually) D.2. D.0 holds trivially. Given $\alpha_1,\ldots,\alpha_n \in E$, $\alpha_1 - y$ is full (over $E_k\langle y\rangle$) but $\alpha_1 - \alpha_1 = 0$.

To prove D.2 we shall assume that k is infinite and K is infinite-dimensional over its centre k. Let $\alpha_1,\ldots,\alpha_n,\beta \in E$

and suppose that β is dependent on $\alpha_1,\ldots,\alpha_n$ but not on $\alpha_2,\ldots,\alpha_n$. By hypothesis there is a matrix A over $K_k\langle x_1,\ldots,x_n,y\rangle$ which is full over $E_k\langle y\rangle$ when x_i is replaced by α_i and goes over into a singular matrix when y is replaced by β. Let us write

$$A(x_1,y) = A(x_1,\alpha_2,\ldots,\alpha_n,y),$$

then $A(\alpha_1,y)$ is full and $A(\alpha_1,\beta)$ is singular. If $A(x_1,\beta)$ is full, this will show that α_1 is dependent on $\beta,\alpha_2,\ldots,\alpha_n$, as required, so assume that $A(x_1,\beta)$ is not full. Clearly $A(x_1,y)$ is full, hence by the specialization lemma there exists $\xi \in K$ such that $A(\xi,y)$ is full. Now $A(\xi,\beta)$ is singular (because $A(x_1,\beta)$ is not full), hence β is algebraically dependent on $\alpha_2,\ldots,\alpha_n$ over K, a contradiction.

If D.1 were true, we could conclude in the usual way (cf. e.g. Cohn [77]) that every extension field of K has a transcendence basis with a uniquely determined cardinal ν ("transcendence degree") and any algebraically independent set has at most ν elements. But in Th. 5.5.6 we saw that every countably generated field can be embedded in a 2-generator field over k, and there is no difficulty in deriving such an embedding over a field K infinite-dimensional over its centre. This shows that no transcendence degree can exist, so D.1 does not hold in general. A look at the proof in the commutative case shows that what is needed is a form of elimination. This again emphasizes the point that in the case of skew fields a general elimination procedure is lacking, but it might be worthwhile to examine special situations where elimination can be used.

List of notations

(standard items such as **N**, **Z**, **Q**, **R**, **C** have not been included)

Bibliography

Numbers in italics refer to page numbers in the book.

Amitsur, S.A.

48. A generalization of a theorem on differential equations, *Bull. Amer. Math. Soc.* 54 (1948) 937-941 *65*

54. Noncommutative cyclic fields, *Duke Math. J.* 21 (1954) 87-105 *48, 65, 70f.*

54'. Differential polynomials and division algebras, *Ann. of Math.* 59 (1954) 245-278

55. Finite subgroups of division rings, *Trans. Amer. Math. Soc.* 80 (1955) 361-386

58. Commutative linear differential operators, *Pacif. J. Mat* 8 (1958) 1-10

65. Generalized polynomial identities and pivotal monomials, *Trans. Amer. Math. Soc.* 114 (1965) 210-226 *141*

66. Rational identities, and applications to algebra and geometry, *J. Algebra* 3 (1966) 304-359 *165, 172*

Asano, K.

49. Über die Quotientenbildung von Schiefringen, *J. Math. So Japan* 1 (1949) 73-78 *8*

Bergman, G.M. *118, 147, 206*

64 A ring primitive on the right but not on the left, *Proc. Amer. Math. Soc.* 15 (1964) 473-475 *25*

67. Commuting elements in free algebras and related topics in ring theory, Thesis Harvard University 1967 *141*

70. Skew fields of noncommutative rational functions, after Amitsur, Sem. Schützenberger-Lentin-Nivat 1969/70 No. 16 (Paris 1970) *166, 168, 171f.*

74. Modules over coproducts of rings, *Trans. Amer. Math. Soc.* 200 (1974) 1-32 *5, 87, 98ff., 103f., 106f., 109*

74'. Coproducts and some universal ring constructions, *Trans. Amer. Math. Soc.* 200 (1974) 33-88

76. Rational relations and rational identities in division rings I, II. *J.Algebra*, 43(1976)252-66, 267-97 *166, 175, 178, 187*

Bergman, G.M. and Small, L.W.

75. PI-degrees and prime ideals, *J. Algebra* 33(1975) 435-462 *174f., 187*

Boffa, M. and v.Praag, P.

72. Sur les corps génériques, *C.R.Acad.Sci. Paris Ser.A* 274 (1972) 1325-1327 *135*

Bokut', L.A.

63. On a problem of Kaplansky, *Sibirsk. Zh. Mat.* 4(1963) 1184-1185 *19*

69. On Mal'cev's problem, *Sibirsk.Zh.Mat.* 10(1969) 965-1005 *3, 91*

Borevič, Z.I.

58. On the fundamental theorem of Galois theory for skew fields, *Leningrad Gos. Ped. Inst. Uč. Zap.*166(1958) 221-226

Bortfeld, R.

59. Ein Satz zur Galoistheorie in Schiefkörpern, *J. reine u.angew. Math.* 201 (1959) 196-206

Bowtell, A.J.

67. On a question of Mal'cev, *J. Algebra* 6 (1967) 126-139 *4, 91*

Brauer, R.

49. On a theorem of H. Cartan, *Bull. Amer.Math.Soc.* 55 (1949) 619-620

Brungs, H.-H.

69. Generalized discrete valuation rings, *Canad. J. Math.* 21 (1969) 1404-1408 *25*

Bryars, D.A. *95*

Burmistrovič, I.E.

63. On the embedding of rings in skew fields, *Sibirsk. Zh. Mat.* 4 (1963) 1235-1240

Cameron, P.J. *117*

Cartan, H.

47. Théorie de Galois pour les corps non-commutatifs, *Ann. Sci. E.N.S.* 64 (1947) 59-77

Cherlin, G.

72. The model companion of a class of structures, *J. Symb. Logic* 37 (1972) 546-556 *134*

Cohn, P.M.

59. On the free product of associative rings, *Math. Zeits.* 71 (1959) 380-398 *98*

60. On the free product of associative rings. II. The case of (skew) fields, *Math. Zeits.* 73 (1960) 433-456

61. On the embedding of rings in skew fields, *Proc. London Math. Soc.* (3) 11 (1961) 511-530 *28*

61'. Quadratic extensions of skew fields, *Proc. London Math. Soc.* (3) 11 (1961) 531-556 *29, 56*

62. Eine Bemerkung über die multiplikative Gruppe eines Körpers, *Arch. Math.* 13 (1962) 344-348

63. Rings with a weak algorithm, *Trans. Amer. Math.Soc.* 109 (1963) 332-356 *106*

64. Free ideal rings, *J. Algebra* 1 (1964) 47-69 *106*

65. Universal Algebra, Harper and Row (New York, London 1965) *4, 20, 91*

66. On a class of binomial extensions, *Ill. J. Math.* 10 (1966) 418-424

66'. Some remarks on the invariant basis property, *Topology* 5 (1966) 215-228 *75*

67. Torsion modules over free ideal rings, *Proc. London Math. Soc.* (3) 17 (1967) 577-599 *25*

68. On the free product of associative rings III, *J. Algebra* 8 (1968) 376-383 *106*

69. Dependence in rings.II. The dependence number, *Trans. Amer. Math. Soc.* 135 (1969) 267-279 *3, 90f.*

71. The embedding of firs in skew fields, *Proc. London Math. Soc.* (3) 23 (1971) 193-213 *91*

71'. Rings of fractions, *Amer. Math. Monthly* 78 (1971) 596-615 *5, 9*

71". Free rings and their relations, *No.2, LMS monographs, Academic Press* (London, New York 1971) *18, 25, 27, 54, 58, 78, 83, 86f., 90, 98, 107, 112, 118, 128f., 141f., 153, 209, 211-216, 221*

71'". Un critère d'immersibilité d'un anneau dans un corps gauche, *C.R.Acad. Sci. Paris, Ser. A*, 272 (1971) 1442-1444

72. Universal skew fields of fractions, *Symposia Math. VIII* (1972) 135-148

72'. Generalized rational identities, *Proc. Park City Conf.* 1971 in Ring Theory (ed. R.Gordon) Acad. Press (New York 1972) 107-115 *147, 166*

72". Skew fields of fractions, and the prime spectrum of a general ring, in Lectures on rings and modules, Lecture Notes in Math. No.246 (Springer, Berlin 1972) 1-71

72"'. Rings of fractions, Univ. of Alberta Lecture Notes 1972

73. Free products of skew fields, *J. Austral. Math. Soc.* 16 (1973) 300-308

73'. The word problem for free fields, *J. symb. Logic* 38 (1973) 309-314, correction and addendum ibid. 40 (1975) 69-74

73". The similarity reduction of matrices over a skew field, *Math. Zeits.* 132 (1973) 151-163 *118, 206*

73"'. The range of derivations on a skew field and the equatic $ax - xb = c$, *J. Indian Math. Soc.* 37 (1973) 1-9 *19*

73^{iv}. Skew field constructions, Carleton Math. Lecture Notes No.7 (Ottawa 1973)

74. Progress in free associative algebras, *Israel J. Math.* 19 (1974) 109-151

74'. Localization in semifirs, *Bull. London Math. Soc.* 6 (1974) 13-20

74". The class of rings embeddable in skew fields, *Bull. London Math. Soc.* 6 (1974) 147-148 *91*

75. Presentations of skew fields. I. Existentially closed skew fields and the Nullstellensatz, *Math. Proc. Camb. Phil. Soc.* 77 (1975) 7-19 *140*

76. The Cayley-Hamilton theorem in skew fields, *Houston J. Math.* 2 (1976) 49-55 *202f.*

76'. Equations dans les corps gauches, *Bull. Soc. Math. Belg.* *201, 223*

77. Algebra vol.2, J. Wiley (London, New York 1977) *42, 54, 162, 173, 230f.*

(a) Zum Begriff der Spezialisierung über Schiefkörpern, to appear

(b) The universal field of fractions of a semifir, to appear *168*

(c) A construction of simple principal ideal domains, to appear *58*

Cohn, P.M. and Dicks, W.

76. Localization in semifirs II, *J. London Math. Soc.* (2) 13 (1976) 411-418 *142*

Dauns, J.

70. Embeddings in division rings, *Trans. Amer. Math. Soc.* 150 (1970) 287-299 *28*

Dicks, W. *96, 99, 165*

Dickson, L.E. *54*

Dieudonné, J.

43. Les déterminants sur un corps noncommutatif, *Bull. Soc. Math.* France 71(1943) 27-45 *199*

52. Les extensions quadratiques des corps noncommutatifs et leurs applications, *Acta Math.* 87 (1952) 175-242

Doneddu, A.

71. Études sur les extensions quadratiques des corps non-commutatifs, *J. Algebra* 18(1971) 529-540

72. Structures géometriques d'extensions finies des corps non-commutatifs, *J. Algebra* 23(1972) 18-34

74. Extensions pseudo-linéaires finies des corps non-commutatifs, *J. Algebra* 28(1974) 57-87

Faith, C.C.

58. On conjugates in division rings, *Canad. J. Math.* 10 (1958) 374-380

Farkas, D.R. and Snider, R.L.

76. K_o and Noetherian group rings, *J. Algebra* 42 (1976) 192-198 *20*

Faudree, J.R.

66. Subgroups of the multiplicative group of a division ring, *Trans. Amer. Math. Soc.* 124 (1966) 41-48
69. Locally finite and solvable subgroups of skew fields, *Proc. Amer. Math. Soc.* 22 (1969) 407-413

Fisher, J.L.

71. Embedding free algebras in skew fields, *Proc. Amer. Math. Soc.* 30 (1971) 453-458 *15*
74. The poset of skew fields generated by a free algebra, *Proc. Amer. Math. Soc.* 42 (1974) 33-35
74'. The category of epic R-fields, *J. Algebra* 28 (1974) 283-290

Fuchs, L.

63. *Partially ordered algebraic systems,* Pergamon (Oxford 1963) *21*

Gelfand, I.M. and Kirillov, A.A.

66. Sur les corps liés aux algèbres enveloppantes des algèbres de Lie, *Publ. Math. IHES* No. 32 (1966) 5-19

Goldie, A.W. *14*

Gordon, B. and Motzkin, T.S.

65. On the zeros of polynomials over division rings, *Trans. Amer. Math. Soc.* 116 (1965) 218-226, Correction ibid. 122 (1966) 547 *54, 165, 207*

Hahn, H.

07. Über die nichtarchimedischen Grössensysteme, *S.-B. Akad. Wiss. Wien IIa* 116 (1907) 601-655 *20*

Hall, M.

59. *The theory of groups*, Macmillan (New York 1959) *22*

Harris, B.

58. Commutators in division rings, *Proc. Amer. Math. Soc.* 9 (1958) 628-630 *19*

Herstein, I.N.

53. Finite subgroups of division rings, *Pacif. J. Math.* 3 (1953) 121-126

56. Conjugates in division rings, *Proc. Amer. Math. Soc.* 7 (1956) 1021-1022

68. *Non-commutative rings* (Carus Monographs; J. Wiley 1968) *162*

Herstein, I.N. and Scott, W.R.

63. Subnormal subgroups of division rings, *Canad. J. Math.* 15 (1963) 80-83

Higman, G.

40. The units of group rings, *Proc. London Math. Soc.* (2) 46 (1940) 231-248 *152*

52. Ordering by divisibility in abstract algebras, *Proc. London Math. Soc.* (3) 2 (1952) 326-336 *20*

61. Subgroups of finitely presented groups, *Proc. Roy.Soc.*, *Ser. A* 262 (1961) 455-475 *141*

Higman, G., Neumann, B.H. and Neumann, H.

49. Embedding theorems for groups, *J. London Math. Soc.* 24 (1949) 247-254 *115*

Hilbert, D.

1896. *Grundlagen der Geometrie*, Teubner (Stuttgart 1896, 10th ed. 1968)

Hirschfeld, J. and Wheeler, W.H.

75. Forcing, Arithmetic, Division rings, *Lecture Notes in Math.* No. 454, Springer (Berlin 1975) *134f., 137*

Hua, L.K.

49. Some properties of a sfield, *Proc. Nat. Acad. Sci.* USA 35(1949) 533-537

50. On the multiplicative group of a sfield, *Science Record Acad.* Sinica 3 (1950) 1-6

Huzurbazar, M.S.

60. The multiplicative group of a division ring, *Doklady Acad. Nauk SSSR* 131 (1960) 1268-1271 = *Soviet Math. Doklady* 1(1960) 433-435

61. On the theory of multiplicative groups of division rings, *Doklady Akad. Nauk SSSR* 137(1961) 42-44 = *Soviet Math. Doklady* 2 (1961) 241-243

Ikeda, M.

62. Schiefkörper unendlichen Ranges über dem Zentrum, *Osaka Math. J.* 14(1962) 135-144

Jacobson, N.

40. The fundamental theorem of Galois theory for quasi-fields, *Ann. of Math.* 41(1940) 1-7

43. Theory of rings, *Amer. Math. Soc.* (Providence 1943)

55. A note on two-dimensional division ring extensions, *Amer. J. Math.* 77 (1955) 593-599

56. Structure of rings, *Amer. Math. Soc.* (Providence 1956, 1964) *25, 32, 40, 42, 124, 162*

62. *Lie algebras*, Interscience (New York and London 1962) *71*

75. PI-algebras, an introduction, *Lecture Notes in Math.* No. 441 Springer (Berlin 1975) *173*

Jategaonkar, A.V.

69. A counter-example in homological algebra and ring theory, *J. Algebra* 12 (1969) 418-440 *23, 27*

69'. Ore domains and free algebras, *Bull. London Math. Soc.* 1(1969) 45-46 *14*

Kaplansky, I.

51. A theorem on division rings, *Canad. J. Math.* 3(1951) 290-292

70. Problems in the theory of rings revisited, *Amer. Math. Monthly* 77 (1970) 445-454 *18*

Klein, A.A.

67. Rings nonembeddable in fields with multiplicative semigroups embeddable in groups, *J. Algebra* 7(1967) 100-125 *4, 91*

69. Necessary conditions for embedding rings into fields, *Trans. Amer. Math. Soc.* 137(1969) 141-151 *5, 81*

70. A note about two properties of matrix rings, *Israel. J. Math.* 8(1970) 90-92 *5*

70'. Three sets of conditions on rings, *Proc. Amer. Math. Soc.* 25(1970) 393-398

72. A remark concerning embeddability of rings into fields, *J. Algebra* 21(1972) 271-274 *5*

Knight, J.T.

70. On epimorphisms of non-commutative rings, *Proc. Camb. Phil. Soc.* 68 (1970) 589-600 *95*

Koethe, G.

31. Schiefkörper unendlichen Ranges über dem Zentrum, *Math. Ann.* 105 (1931) 15-39

Koševoi, E.G.

70. On certain associative algebras with transcendental relations, *Algebra i Logika* 9, No. 5 (1970) 520-529 *14*

Laugwitz, D.

(a) Tullio Levi-Civita's work on non-archimedean structures, to appear in Levi-Civita memorial volume.

Lazerson, E.E.

61. Onto inner derivations in division rings, *Bull. Amer. Math. Soc.* 67(1961), cf. *Zentralblatt f. Math.* 104 (1964) 33-34 *19*

Leavitt, W.G.

57. Modules without invariant basis number, *Proc. Amer. Math. Soc.* 8(1957) 322-328 *75*

Lenstra, jr., W.H.

74. *Lectures on Euclidean rings,* Bielefeld 1974 *25*

Lewin, J. and Lewin, T.

(a) An embedding of the group algebra of a torsion free one relator group in a field, to appear in *J. Algebra* *20*

Likhtman, A.I.

63. On the normal subgroups of the multiplicative group of a division ring, *Doklady Akad. Nauk SSSR* 152(1963) 812-815 = *Soviet Math. Doklady* 4(1963) 1425

Lyapin, E.S.

60. Semigroups (Moscow 1960, translated *Amer. Math. Soc.* 1963) *23*

Macintyre, A. *115, 123, 158*

73. The word problem for division rings, *J. symb. Logic* 38(1973) 428-436 *158*

(a) On algebraically closed division rings, *Ann. Math. Logic* *135*

(b) Combinatorial problems for skew fields. I Analogue of Britton's lemma, and results of Adyan-Rabin type, to appear

Magnus, W. *20*

37. Über Beziehungen zwischen höheren Kommutatoren, *J. reine u. angew. Math.* 177 (1937) 105-115

Makar-Limanov, L.G.

75. On algebras with one relation, *Uspekhi Mat. Nauk* 30, No.2 (182)(1975) 217 *200*

77. To appear in *Algebra i Logika* *200*

Mal'cev, A.I.

37. On the immersion of an algebraic ring into a field, *Math. Ann.* 113(1937) 686-691 *1*

39. Über die Einbettung von assoziativen Systemen in Gruppen I, II (Russian, German summary), *Mat. Sbornik N.S.* 6 (48) (1939) 331-336 ibid. 8(50) (1940) 251-264 *4*

48. On the embedding of group algebras in division algebras *Doklady Akad. Nauk SSSR* 60 (1948) 1499-1501 *20*

73. *Algebraic systems,* Springer (Berlin 1973) *90f.*

Moufang, R.

37. Einige Untersuchungen über geordnete Schiefkörper, *J. reine u. angew. Math.* 176(1937) 203-223 *22*

Nagahara, T. and Tominaga, H.

55. A note on Galois theory of division rings of infinite degree, *Proc. Japan Acad.* 31(1955) 655-658

56. On Galois theory of division rings I,II, *Math. J. Okayama Univ.* 6 (1956) 1-21, ibid. 7 (1957) 169-172

Nakayama, T.

53. On the commutativity of certain division rings, *Canad. J. Math.* 5 (1953) 242-244

Neumann, B.H.

49. On ordered groups, *Amer. J. Math.* 71 (1949) 1-18

49'. On ordered division rings, *Trans. Amer. Math. Soc.* 66 (1949) 202-252 *20*

54. An essay on free products of groups with amalgamations, *Phil. Trans. Roy. Soc. Ser. A* 246 (1954) 503-554 *94, 110, 120*

73. The isomorphism problem for algebraically closed groups, in *Word problems* ed. W.W. Boone et al. North Holland (Amsterdam 1973) *137*

Neumann, H.

48. Generalized free products with amalgamated subgroups. I. Definitions and general properties, *Amer. J. Math.* 70 (1948) 590-625

49. Generalized free products. II. The subgroups of generalized free products, *Amer. J. Math.* 71 (1949) 491-540

Nivat, M.

70. Séries rationnelles et algébriques en variables non

commutatives, *Cours du DEA 1969/70* *78*

Niven, I.

41. Equations in quaternions, *Amer. Math. Monthly* 48 (1941) 654-661

Noether, E. *8, 46*

Ore, O.

31. Linear equations in non-commutative fields, *Ann. of Math.* 32(1931) 463-477 *8*
32. Formale Theorie der linearen Differentialgleichungen, *J. reine u. angew. Math.* 167 (1932) 221-234, ibid. 168 (1932) 233-252
33. Theory of non-commutative polynomials, *Ann. of Math.* 34 (1933) 480-508

v.Praag, P. *201*

71. Groupes multiplicatifs des corps, *Bull. Soc. Math. Belg.* 23(1971) 506-512

Procesi, C.

68. Sulle identità delle algebre semplici, *Rend. Circ. Mat. Palermo ser 2, XVII* (1968) 13-18 *163, 165*
73. *Rings with polynomial identities*, (M. Dekker, New York 1973) *165*

Richardson, A.R.

27. Equations over a division algebra, *Mess. Math.* 57 (1927) 1-6

Robinson, A.

71. On the notion of algebraic closedness for non-commutative groups and fields, *J. symb. Logic* 36 (1971) 441-444 *134*

Sacks, G.

72. *Saturated model theory*, Benjamin (New York 1972) *18*

Schenkman, E.V.

58. Some remarks on the multiplicative group of a sfield, *Proc. Amer. Math. Soc.* 9 (1958) 231-235

61. Roots of centre elements of division rings, *J. London Math. Soc.* 36 (1961) 393-398

Schreier, O.

27. Über die Untergruppen der freien Gruppen, *Abh. Math. Sem.* Hamburg 5 (1927) 161-183 *93*

Schützenberger, M.P.

62. On a theorem of Jungen, *Proc. Amer. Math. Soc.* 13 (1962) 885-890 *78*

Scott, W.R.

57. On the multiplicative group of a division ring, *Proc. Amer. Math. Soc.* 8 (1957) 303-305

Shelah, S.

73. Differentially closed fields, *Israel J. Math.* 16 (1973) 314-328 *18*

Šimbireva, E.P.

47. On the theory of partially ordered groups, *Mat. Sbornik* 20 (1947) 145-178

Sizer, W.S.

75. *Similarity of sets of matrices over a skew field*, Thesis (London University 1975) *215*

77. Triangularizing semigroups of matrices over a skew field, Linear Algebra and its applications, to appear *206*

Small, L.W.

65. An example in Noetherian rings, *Proc. Nat. Acad. Sci. USA* 54 (1965) 1035-1036 *25*

Smith, D.B.

70. On the number of finitely generated O-groups, *Pacif. J. Math.* 35 (1970) 499-502 *123*

Smits, T.H.M.

68. Skew polynomial rings, *Indag. Math.* 30 (1968) 209-224 *18*

Sweedler, M.E.

75. The predual theorem to the Jacobson-Bourbaki correspondence, *Trans. Amer. Math. Soc.* 213 (1975) 391-406 *32*

Szele, T.

52. On ordered skew fields, *Proc. Amer. Math. Soc.* 3 (1952) 410-413

Treur, J.

76. *A duality for skew field extensions,* Thesis (Utrecht 1976)

v.d. Waerden, B.L.

48. Free products of groups, *Amer. J. Math.* 70 (1948) 527-528 *94*

Wähling, H.

74. Bericht über Fastkörper, *Jber. Deutsch. Math.-Ver.* 76 (1974) 41-103

Wedderburn, J.H.M.

09. A theorem on finite algebras, *Trans. Amer. Math. Soc.* 6 (1909) 349-352

Wiegmann, N.A.

55. Some theorems on matrices with real quaternion elements, *Canad. J. Math.* 7 (1955) 191-201 *217*

Wolf, L.A.

36. Similarity of matrices in which the elements are real quaternions, *Bull. Amer. Math. Soc.* 42 (1936) 737-743

Index